LICE

Written and Edited by: Dr Austin Mardon, Hadia Saleem, Ezzah Inayat, Faith Dong, Diana Amiscaray, Ellen Mak, Jessica Henry, Minahil Syed, Leah Heinen, Ipsa Gusain, Grace Parish

First Printing: 2021

Typeset and Cover Design by CJ Harrison

ISBN: 9781008951655

Copyright May 2021, forthcoming

Lice
Golden Meteorite Press
103 11919 82 St NW
Edmonton, AB T5B 2W3
www.goldenmeteoritepress.com

Contents

Evolution, lice throughout history.

Hadia Saleem

Louse (plural: lice) is a common name for members of the order Phthiraptera, which contains approximately 5,000 species of insects (Amanzougaghene et al., 2020). According to historical evidence, human lice infestations were present throughout the world for thousands of years. Genetically, human lice are distributed into three mitochondrial clades (A, B and C) (Amanzougaghene et al., 2020). Clade A comprises head and body lice and is found in all continents (Amanzougaghene et al., 2020). Clade B is known to have an American origin and is present in Europe, Australia and North Africa (Amanzougaghene et al., 2020). Finally, Clade C is limited to Africa and Asia (Amanzougaghene et al., 2020). Lice are among the oldest parasites of humans, making them an excellent marker of the evolution and migration of Homo species over time(Amanzougaghene et al., 2020). Additionally, due to the long-term association of lice and humans, lice have become a model for the study of co-phylogenetic relationships between hosts and parasites (Boutellis et al., 2014). Currently, several million children worldwide are infested with head lice annually (Amanzougaghene et al., 2020). Lice infestations are rampant among school children in both developing and developed countries, and across all socioeconomic levels (Amanzougaghene et al., 2020). This chapter will explore the historical roots of lice, by first exploring the taxonomy and morphology of lice (to familiarize the reader with the terminology used in later parts of the chapter). This will then be followed by a specific focus on the evolutionary history of lice and their application to modern-day developments.

Taxonomy of Lice

Lice are small, wingless insects that are unable to live independently from their host. Traditionally, the order of lice is Phthiraptera and is divided into two major groups, Anoplura (sucking lice of placental mammals) and Mallophaga (chewing lice of birds and placental mammals) (Boyd & Reed, 2012). However, recent classifications are now divided into four suborders under the order Phthiraptera, which include Anoplura (sucking lice), Amblycera (primitive suborder of chewing lice), Ischnocera (avian chewing lice), and Rhynchopthirina (parasites of elephants and warthogs) (Boyd & Reed, 2012). Scientific evidence suggests that sucking and chewing lice originate from a common non-parasitic ancestral group related to the order Psocodea (booklice and bark lice) (Bonilla et al., 2013). These groups diverged in the late Jurassic or early Cretaceous period, approximately 100-150 million years ago (Boutellis et al., 2014). There are 540 described species of sucking lice that occupy twelve mammalian orders (Bonilla et al., 2013). Humans can be infested with two types of sucking lice: Pediculus humanus on the head and/or body and Pthirus pubis in the pubic area (Bonilla et al., 2013). Pediculus humaus belongs to the genus of Pediculus, while Pthirus pubis belongs to the genus Pthirus (Bonilla et al., 2013). Additionally, the Pediculus humanus further comprises two ecotypes: Pediculus humanus capitis (head lice) and Pediculus humanus humanus (body lice) (Bonilla et al., 2013). The diet of suckling lice is strictly received by piercing the skin of their human hosts and feeding on their blood (Bonilla et al., 2013). On the other hand, there are 4500 recognized species of chewing lice found on both mammals and birds (Boyd & Reed, 2012). The diet of chewing lice is mostly composed of keratin-rich dermal components such as feathers, skin and hair (Boyd & Reed, 2012). Many birds and mammals can harbour multiple species of louse, and in some cases, different louse specimens may be restricted to only one region of their host (Boyd & Reed, 2012).

Morphology and Biology

Due to the variety of head and body lice, on the same individual, multiple subspecies of lice have been identified. Also, there are no consistent points of distinctive morphology of lice; identification relies on the location of the lice on the human body.

Sucking lice range in length from 0.5 to 5 millimetres, with narrow and oval heads, and flattened bodies (Capinera, 2008). Their antennae are relatively short, consisting of three to five segments (Capinera, 2008). Sucking lice have mouth parts that are adapted to piercing and sucking. Their mouthparts also consist of toothed proboscis, containing a salivary and food canal (Capinera, 2008). Finally sucking lice have fused thoracic segments, separated abdominal segments, and a single large claw at the tip of each individual leg (six in total) (Capinera, 2008). The two ecotypes of sucking lice have similar morphology but differ in their ecology and nutritional patterns. Head lice live, breed, and lay their eggs (nits) at the base of hair shafts and feed on human blood every four to six hours (Bonilla et al., 2013). Body lice live and lay their eggs in clothes, feed less frequently and overall ingest large quantities of blood compared to head lice (Bonilla et al., 2013). Additionally, body lice are more resistant to environmental conditions, resulting in them having a more than seventy-two hours off-host survival rate (Bonilla et al., 2013). According to current research, it is understood that lice found on the human body, clothing and other personal effects such as bedding have evolved specific behavioural and physiological adaptations that are not shared by head lice (Boutellis et al., 2014). For instance, the average size of head lice is smaller than that of body lice, the antennae of the two show considerable differences in proportions and the lateral indentations between segments of the abdomen are more pronounced in typical head lice than in body lice (Boutellis et al., 2014). Though the colour of lice can vary from pale beige to dark gray, when feeding on blood, it can become considerably darker (Boutellis et al., 2014). Head lice are generally darker than body lice (Boutellis et al., 2014).

Chewing lice also possess flattened bodies, but have slightly wider heads (Capinera, 2008). They are also larger than sucking lice, ranging from 0.5 to 6 mm (Capinera, 2008). Chewing lice have no ocelli and their mouthparts are adapted for chewing (Capinera, 2008). The antennae of chewing lice have three to five segments, and their legs are short, ending with one or two claws per leg (Capinera, 2008). Nonetheless, it is necessary to note that the difference between the various types of lice is largely dependent on the environment it resides in.

Evolutionary History of Lice

Lice are known for showing more cospeciation with hosts than other groups of insects (Boutellis et al., 2014). The oldest human head louse was found on a hair from an archaeological site in northeastern Brazil and was dated to 8000 B.C (Boutellis et al., 2014). Head lice have also been found at archeological sites in the southwestern USA, the Aleutian Islands, Peru, Greenland, Mexico and on mummies that were Incan sacrifices. Another discovery of lice was reported from a Maitas Chiribaya mummy from Africa, in Northern Chile recently (Boutellis et al., 2014). This discovery was dated back to 670-900 years ago. The Psocoptera order (including book lice) is hypothesized to have originated in the Mesozoic Era (Boutellis et al., 2014). Due to the exceptional circumstances required for the fossilization of lice, and the scarcity of the fossils themselves, the origin and evolution of lice are highly uncertain (Boutellis et al., 2014). There has also been significant debate among researchers over the exact age of lice and the origins of parasitism in this group. The recent discovery of two important fossils: Saurodectes vrsanskyi and Megamenopon rasnitsyni provided researchers with valuable information on the age of lice (Boutellis et al., 2014). Saurodectes vrsanskyi was recovered from the Zaza formation shales of Bassia, Siberia (Boutellis et al., 2014). This fossil is approximately ten times larger than any currently living louse. Researchers believe that this louse possibly resided on a massive host (Boutellis et al., 2014). The Megamenopon rasnitsyni, a well-preserved louse fossil, was found in the oil shale of Eckfeld Maar. it is twice the length of similar present-day lice (Boutellis et al., 2014).

Humans (Homo sapiens) are parasitized by suckling lice, classified as Anoplora, which depending on the subspecies, belong to the Pediculus and Pthirus genus. The three subspecies include head lice (Pediculus humanus capitis), body lice (Pediculus humanus humanus), and pubic lice (Pthirus pubis) (Reed et al., 2007). Interestingly, the closest living relatives to humans, chimpanzees and gorillas harbour Pediculus and Phthirus species respectively (Reed et al., 2007). This means that humans share a genus with both chimpanzees and gorillas. Researchers propose that Approximately six million years ago, humans and apes evolutionarily separated, splitting the Pediculus genus (Reed et al., 2007). This resulted in the Pediculus schaeffi species becoming a part of the chimpanzee lineage, while the Pediculus humanus species became a part of the homo lineage (Reed et al., 2007). Similarly, the gorillas are hypothesized to be split off one million years earlier than the apes. This evolutionary split resulted in the Pthirus gorillae joining the gorilla's lineage, while Pthirus pubis, joining the homo lineage (Reed et al., 2007). Though the exact process through which humans may have acquired pubic lice is unknown, researchers are very interested in discovering the timeline in which this switch first occurred, and whether it was recent or considerably older (Reed et al. 2007). Researchers have published multiple hypotheses to explain why humans carry chimpanzee and/or gorilla-related genus of lice, of which the most common two being the 'pair of lice lost' model and the 'recent host switch' theory.

Pair of Lice Lost Model

The 'pair of lice lost' model requires three evolutionary steps: one duplication of the parasite and two extinctions of the Pediculus and Pthirus lineage (Reed et al., 2007). In this theory, cophylogenetic reconstruction is composed of an ancient duplication, creating two evolutionarily distinct lineages (Pediculus and Pthirus) (Reed et al., 2007). Each of these lineages cospeciated with gorillas, chimpanzees, and humans, with two extinction events (Reed et al., 2007). This theory suggested that after apes and humans parted ways, each lineage carried both Pediculus and Pthirus genus. Along this process, chimpanzees dropped Pthirus and gorillas dropped Pediculus, while humans ended up inheriting both.

Recent Host Switch Theory

The 'recent host switch' hypothesis requires one evolutionary step and predicts that the divergence between Pthirus pubis and Pthirus gorillae is more recent than the chimpanzee/human split (Reed et al., 2007). In this theory, cophylogenetic reconstruction is composed of perfect co-speciation between the hosts and parasites, with the exception of a single host switch of Pthirus species from gorillas to humans (Reed et al., 2007). This theory suggests that hosts were switched as a result of direct contact from gorillas to humans. Although human pubic lice are most commonly transmitted through sexual contact, such contact may not be required to explain the host switch (Reed et al., 2007). Transfer of lice from gorillas and chimpanzees could also possibly be due to predation, overlaps in habitats, and/or natural changes in the homo lineage (Reed et al., 2007). The host switch could have also been possible by any form of contact between humans and gorillas including feeding on or living among gorillas (Reed et al., 2007). Nevertheless, regardless of the method of parasite transfer, an adequate habitat had to be available on the new human host in order for the host switch to be successful. For example, the switch of Pthirus from gorillas to humans could have possibly coincided with a change in available niche space in humans, such as loss of body hair.

The aforementioned hypotheses on the evolutionary history of lice are based on the two primary premises, (1) to maximize the number of cospeciation events and (2) to minimize the number of events that deviate from cospeciation (Boutellis et al., 2014). The 'pair of lice lost' model is considered to be less parsimonious than the 'recent host switch' hypothesis because it requires more evolutionary steps (Boutellis et al., 2014). Nevertheless, it is important to note that this does not make one hypothesis more likely to occur compared to the other; all historical events (host switch, duplication and extinction) have some probability to occur (Boutellis et al., 2014). Additional hypotheses with more evolutionary steps than the 'pair of lice lost' model and 'recent host switch' theory do occur, however, specific research is limited (Boutellis et al., 2014). In terms of the separation of head and body lice, according to the mtDNA (mitochondrial DNA) molecular cloak analysis from a global sample of forty head and body lice, the separation of Pediculus Humanus clade A into distinct "head" and "body" lice is estimated to have occurred approximately $72,000 \pm 42,000$ years ago (Boutellis et al., 2014). In terms of the gene content and arrangement of the mitochondrial minichromosomes of head and body lice, they are identical. This indicates that despite the type of lice separated, the characteristics of minichromosomes have not changed (Boutellis et al., 2014). Moreover, although there is a theory that suggested that body lice evolved from head lice, it is not completely consistent with molecular data (Boutellis et al., 2014).

Application of Research on the Evolutionary History of Lice

The evolutionary history of lice facilitates a better understanding of our past, which can be used to explain more recent developments. Genetic evidence and DNA differences between head lice and body lice also provide researchers with evidence that humans used clothing approximately 80,000 and 170,000 years ago, prior to leaving Africa (Toups et al., 2010). Since body lice required clothing to survive, the divergence of head and body lice from their common ancestor suggests an estimated date of clothing in human evolutionary history (Toups et al., 2010). Additionally, research on the evolutionary history of lice sheds light on human migratory patterns in prehistory. An analysis of the mitochondrial DNA in human and body lice revealed that African lice carried greater genetic diversity compared to non-African lice (Kittler et al., 2004). Most anthropologists follow the Out of Africa Hypothesis (evolutionary theory of modern human origin that posits that modern humans arose in the late Pleistocene, about 100,000 to 200,000 years ago, in Africa) regarding human migration (Kittler et al., 2004). Since there is more genetic diversity in African lice, this suggests that the lice and their human hosts must have existed in Africa before moving to anywhere else in the world. Additionally, the mitochondrial genome of the human species of body lice, head lice, and pubic lice disintegrated into separate microchromosomes, at least seven million years ago (Boutellis et al., 2014).

Mummies of various origins have significantly contributed to information regarding the evolution of human lice. They have helped researchers determine three separate lineages of Pediculus (one distributed globally, and the remaining two only in certain regions) and what human ancestors had to deal with (Raoult et al., 2008). For instance, one Peruvian mummy dating back to 1025 CE was infested with over 406 head lice (Raoult et al., 2008). Additionally, according to some research evidence, mummy specimens dating back to 2000 years were culled from hair combs (Raoult et al., 2008). In 2002, researchers also found pubic lice attached to pleats of clothes from 1000-year-old Peruvian mummy (Raoult et al., 2008)

Conclusion and Chapter Summary

In conclusion, current research suggests that lice have been sucking primate blood for the past 25 million years (Reed et al. 2007). In the twenty-first century, human lice infestation remains widespread all over the world. New discoveries into the biology, epidemiology and evolutionary history of lice, have stimulated a renewed interest in all forms of lice. Future studies should continue to put theoretical work forward into practice when discovering different forms of lice. This will help discover and learn more about the complex interactions between hosts and parasites, as well as facilitate effective control methods for diseases that can spread through human populations.

How were lice discovered? How were treatments discovered?

Ezzah Inayat

Lice have plagued the human race for thousands of years. In fact, lice are older than modern humans altogether! Using mitochondrial and nuclear DNA, scientists recently determined that human and ape Pediculus lice diverged 5.5 million years ago. According to their data, this occurred around the same time as the divergence of the ancestors of modern humans from Homo erectus (Weiss, 2009). This poses the question: when were lice discovered? The answer to this question is not a simple one. Humans have lived with lice for centuries, however scientists have only recently been able to gain knowledge on the characteristics, species, and treatment of lice. This was only made possible because of the knowledge contributed by physicians, scientists, and professionals over hundreds of years. In this chapter, we will discuss the discovery of lice, attempting to piece together the contributions that different individuals from different time periods made to our current knowledge. From the mention of lice in religious scriptures and ancient Egypt, to the start of formal documentation of lice in the 17th century, this chapter will cover the historical developments made in the discovery of lice. In addition, we will discuss how different treatments of lice were discovered over the years, and how these treatment methods compare to modern methods used today.

The Discovery of Lice:

Although formal documentation of lice began in the 16th-17th century, the discovery of lice can be traced back to ancient times. There is even mention of lice in religious scriptures! The Bible refers to lice as the 'third plague' that visited the Egyptians when the Pharaoh denied Moses' request to let his people go (Mumcuoglu, 2008). This religious text, along with ancient Sumerian, Akkadian, and Egyption sources provide evidence that those who lived in the Middle East in ancient times were familiar with lice (Mumcuoglu, 2008). For this reason, it is difficult to attribute the discovery of lice to a specific individual or time period, rather it can be stated that these civilizations each independently discovered lice through their own means. Knowledge of lice was not limited to the Middle East. There is strong evidence that suggests many civilizations outside of the Middle East were also well acquainted with lice. For example, lice were identified on ancient peruvian mummies, as well as on the scalps of pre-Columbian Native Americans (Weiss, 2009). Louse combs that date back to Pharaonic times in Egypt also confirm that humans were well aware of the presence of lice, and made efforts to treat lice infestations with the resources available to them at the time (Mumcuoglu, 2008).

For hundreds of years, humans lived aware of the presence of lice, however lice were not defined or documented until the 16th/17th century. Francesco Redi, an Italian physician, was the first to classify lice as a distinct species through his description of P. pubis in 1668 (Stanford Education, n.d.). P. pubis, also known as crab or pubic lice, are typically found in the pubic hair of humans and can spread through sexual contact (Centre for Disease Control and Prevention, 2019). Shortly after, Carl de Geer described P. capitis (head lice). In the 18th century, Joseph Jakob Plenck first described the term Pediculosis, as the infestation of lice (Stanford Education, n.d.). He also described five different types of lice and the areas of the body they affect. In 1842, a surgeon by the name of Erasmus Wilson defined lice and Pediculosis using modern terminology in a textbook he published (Stanford Education, n.d.). Since then, many physicians and scientists have contributed to our current knowledge of lice through experimentation and research. New species, characteristics, and terminology were discovered over the years, and this new knowledge has aided in the development of multiple treatment methods. One of the most prominent and early contributors to the modern understanding of lice was Hans Zinsser. Zinsser was born in New York in 1878, and soon became a world renowned bacteriologist and epidemiologist (The Editors of the Encyclopedia Britannica, 2009). In 1935, his book titled Rats, Lice, and History, was published. This book immediately captured the interest of many, and inspired other scientists to conduct their own research on lice. As a result, Rats, Lice, and History became the foundation of modern day knowledge of lice. And so, after thousands of years, humans know more about lice than we ever have before! It is clear that credit for the discovery of lice cannot be given to a single individual, rather lice were discovered independently in multiple ancient civilizations, after which scientists, physicians, and scholars built upon this discovery through thorough experimentation and research.

The Discovery of Treatment:

The infestation of lice, otherwise known as Pediculosis, has plagued the human race for centuries. Although lice do not present any severe threat to human health, Pediculosis can be an extremely uncomfortable experience for children and adults alike. The most common symptom of Pediculosis is itching, also known as pruritus. Scratching can further lead to the formation of sores on the scalp, which can potentially cause a bacterial infection (Centre for Disease Control and Prevention, 2019). Given this information, it is clear that experiencing Pediculosis can be uncomfortable. This demonstrates why there was a need for the development of appropriate treatments for lice. Throughout hundreds of years, humans have developed multiple ways of treating lice. These treatment methods originate from different time periods and individuals, which is why similar to the discovery of lice, credit for the discovery of treatment cannot be given to a single individual. Over the years, treatment methods have changed due to the advancement of society.

Historical Methods of Treatment:

The earliest method of treating lice was with the hands. This was practiced not only by humans, but by animals as well. In fact, many speculate that humans discovered this technique of removing lice through observing apes. Apes frequently engage in grooming and nit-picking behaviours. Not only does this behaviour allow for the elimination of lice, but it promotes social bonding (Weiss, 2009). In ancient times, humans with Pediculosis would also use their fingers to remove and crush lice from the scalp (Sangaré, 2016). This was often done by another member of the community. Although this treatment method was successfully used in ancient times, it is highly discouraged today. Manual removal of lice can lead to bacterial infection, and is not considered to be as effective as modern day methods of treatment. After years of relying on their hands and eyesight for the removal of lice, ancient civilizations developed louse combs. The use of combs to remove lice is still practiced today in many cultures. The process is quite simple. An individual must comb each section of their hair, examining the comb for lice between sections and rinsing it. This can be done every 1-3 days, and is extremely useful as both a treatment method and preventative measure for combating lice (Sangaré, 2016). In addition to the use of louse combs, shaving soon became a popular method of treating lice. This treatment method was a simple and successful way to eliminate lice and prevent reinfestation, and was very popular in ancient Egypt. Many are aware that ancient Egyptians used to shave their heads, and wore decorative wigs and headdresses as a symbol of power, however not many people are aware that this was a way of preventing lice (Lice Clinics of Texas, 2019). Despite its popularity in ancient Egyptian civilizations, this method of treatment is not commonly practiced in the modern world. Treatment of lice with the hands, louse combs, and shaving were all treatment methods that were discovered in ancient times, hence why there is no documented evidence of a single individual or region introducing these treatment methods to society. As time progressed, new methods of treatment were discovered to be effective at eliminating lice. One of these methods included the use of heat. Those who suspected a lice infestation would apply hot water to their clothing and bedding in an attempt to destroy lice (Sangaré, 2016). In 1917, a man by the name of Arthur William Baccot explained in his work 'The Louse Problem' that a temperature of 52 degrees celsius for a total of thirty minutes is sufficient to destroy lice and lice eggs (Bacot, 1917). Citizens of England used this knowledge to eliminate lice from their clothing and bedding, however this was not a feasible method of treatment for infested hair. As a result, many researchers began to investigate the use of chemicals for the treatment of lice in an attempt to discover an effective method to combat Pediculosis. Sulfur, mercury, and petroleum are only a few of the chemicals that were studied and used to treat lice, however after adverse reactions and resistance, these methods of treatment were quickly discontinued and are no longer used by the general population (Sangaré, 2016).

Insecticides and Medicinal Compounds:

After the second world war, the use of insecticides for the treatment of lice became quite popular. Organochlorines DDT and lindane were among the first of these insecticides. DDT was developed in the 1940's and quickly became one of the most popular treatment methods. It was often used to treat war prisoners for lice, and soon gained popularity with the general population as well (Sangaré, 2016). Lindane was made available in 1951, and also quickly became a popular method of treating lice. Malathion is another insecticide that was commonly used in the United States and Europe for treating lice. It proved to be extremely effective, however scientists began to find cases of Malathion resistance in many countries. In addition, this insecticide is a fire hazard and contributed to chemical burns, which soon resulted in a decline in its popularity (Sangaré, 2016). In 2014, scientists discovered that the use of a 4% dimethicone gel was effective against lice (Sangaré, 2016). There have been many clinical trials conducted using dimethicone against lice, most of which have produced positive results. Benzyl Alcohol is another substance that was used to treat Pediculosis given its ability to kill lice via asphyxiation. This substance however, is associated with multiple negative side effects, including irritation of the eyes, bacterial infection, and inflammation/irritation of the skin. For this reason, it is no longer used for the treatment of lice in the United States, and requires a prescription to obtain in Canada (Sangaré, 2016). In 1997, a company called DowAgroSciences created a product called Spinosad which they produced using the Saccharopolyspora spinosa bacteria found in soil (Salgado, 1999). A 0.9% topical solution of Spinosad proved to be effective in the treatment of lice for many years, and is still used to treat lice today (Sangaré, 2016). These are only a few of the products used throughout history to treat lice. Each of these products was developed in a different time period by different individuals and companies, however many of these products are still useful in the treatment of lice today.

New Treatment Methods and Upcoming Discoveries:

There are a wide variety of treatment options available to combat lice today, however scientists are constantly on the hunt for new and more effective treatment methods. Scientists have recently begun to investigate the symbiotic treatment of lice. Endosymbionts can be referred to as organisms that have a symbiotic relationship with another organism (Casem, 2016). A symbiotic relationship is a mutually beneficial relationship between two organisms. Take coral for example, where the multicellular cnidarian and unicellular dinoflagellate have a symbiotic relationship. The cnidarian is the host, and provides protection and nutrients such as nitrogen, phosphorus, and sulfur to the endosymbiont dinoflagellate. In return, the dinoflagellate provides the host with glucose, resulting in both parties benefiting from the interaction (Casem, 2016). Using their knowledge of symbiotic relationships, many scientists are working to discover new treatments for lice that utilize symbiotic relationships. Using an in vitro feeding model, scientists recently discovered that administration of the antibiotic doxycycline given at varying dosages for ten days can affect the endosymbionts of lice and decrease lice egg production (Sangaré, 2016). This is a huge breakthrough in the scientific community and may contribute to the discovery of a new treatment method for lice. Scientists are also investigating the use of other antibiotics to treat lice such as erythromycin and azithromycin. In addition, scientists are looking into the combination of various drugs to treat lice, rather than treating them independently with each drug (Sangaré, 2016). In the coming years, scientific breakthroughs may allow for a complete change in the way we treat lice.

Barriers to Research

Scientists continue to research lice in an effort to learn more about these parasitic insects and their future implications for society. Ongoing research is crucial for the development of new treatments for lice infestations and treatments for diseases caused by lice. Lice research may also help scientists gain knowledge on other parasitic insects, which is why this type of research should be prioritized. Despite its importance, there are some barriers to this research, the most predominant barrier being a lack of funding. Given the fact that lice do not pose a severe threat to human health, there is less incentive to allocate funds for research on lice in the future. This poses a barrier to the development of new treatments for lice and lice-related diseases.

Chapter Summary:

The discovery of lice can be traced back to ancient times. Typically, a discovery is associated with an individual or group of individuals, however this is not true in the case of lice. Multiple civilizations independently discovered lice and came up with appropriate treatment methods given the resources available to them at the time. There is sufficient evidence that inhabitants of the Middle East and other ancient civilizations were well acquainted with lice though these inhabitants did not have as much knowledge as humans have today. In fact, it was not until the 16th-17th century that lice began to be documented and classified as a distinct species. This was the result of the hard work of many important scientists and physicians, including Francesco Redi, Carl de Geer, Joseph Jakob Plenck, Erasmus Wilson, and Hans Zinsser. Similar to the discovery of lice, the treatment of lice was gradually developed and built on by many individuals. Historical methods of treatment included the mimicking of apes and other animals in using the hands to remove lice, the use of louse combs, shaving of the head and body, as well as the application of heat. These methods were used for many years and proved to be successful, however with the advancement of modern day science came the discovery of insecticides and other medicinal compounds that were used to treat lice, such as DDT, lindane, and Benzyl Alcohol. The treatment of lice has advanced in many ways, however scientists are constantly looking to expand human knowledge on lice and discover new methods of treatment. Recent studies dealing with the use of symbiotic treatment and a combination of drugs/antibiotics for the treatment of lice have been gaining traction, and may change the way in which humans treat lice in the near future.

What is the impact of lice?

Faith Dong

Lice can have widespread impacts on society. For example, lice are known to transmit potential life-threatening diseases. In particular, body lice can act as a targeted vehicle for deleterious pathogens - such as the bacteria Rickettsia prowazekii, Bartonella quintana, and Borrelia recurrentis - which are responsible for many of these human transmitted illnesses. Furthermore, the transmission of disease through lice can lead to significant impacts on society such as mass outbreaks, which can cause short term and long term consequences in the social structure and economy of a society.

The focus of this chapter will be the exploration of the various diseases that are transmitted by body lice, as well as the macroscale impacts on society that they entail. Such diseases or illnesses include epidemic typhus, trench fever, and epidemic relapsing fever, all of which heavily impact communities with high infection rates. Past outbreaks of diseases transmitted by body lice will also be discussed in this chapter, with an emphasis on outbreaks in armies and troops during World War I and World War II. Finally, this chapter will also investigate the continuing impact of lice in today's society as they pose a risk for the resurgence of these diseases through their connection with detrimental pathogens.

Epidemic Typhus

Epidemic typhus is a disease that is transmitted through body lice, and has proven to cause significant social and economic devastation to society in World War I and World War II. The impact of epidemic typhus also continues to this day, particularly in regions where the living conditions are poor, unhygienic, and overcrowded. Symptoms of epidemic typhus often begin within two weeks after contact with infected body lice. The symptoms may range from chills, body and muscle aches, coughing, nausea, vomiting, and confusion (Centers for Disease Control and Prevention). This disease is also characterized by signs such as severe headaches, fever, severe myalgias, arthralgias, rashes, and more (Bechach, Capo, Mege, & Raoult, 2008). These symptoms may range in severity, depending on the age group and general health of the affected individual.

Transmission of Epidemic Typhus

Charles Nicolle first discovered that epidemic typhus was transmitted through body louse in 1909, where he observed that patients in the hospital that underwent bathing and a change of clothes were no longer contagious. Moreover, he observed that the transmission of the disease occurred through close physical contact between hosts (Bechach, Capo, Mege, & Raoult, 2008).

Further analysis revealed that the bacteria Rickettsia prowazekii is the underlying cause of epidemic typhus, and is transmitted to the human body through the body lice and the feces of body lice (Kollipara & Tyring, 2017). Essentially, Rickettsia prowazekii proliferate in body louse, specifically in their gut epithelium, which ultimately kills the body louse through the explosion of Rickettsia prowazekii into the gastrointestinal tract (Akram, Ladd & King, 2021). Thus, the transmission of epidemic typhus occurs through the contamination of the louse bite sites by the pathogen Rickettsia prowazekii (Raoult, Woodward, & Dumler, 2004).

The transmission of epidemic typhus through body louse is an important health consideration for society, even in the present day; if a person with an infection of the bacteria Rickettsia prowazekii is simultaneously infected with lice, infection and large scale transmission can occur in a particular community and affect the given population. Moreover, people who are infected with epidemic typhus retain that bacteria for the rest of their lives, and this could manifest in recurrent form of epidemic typhus, named the Brill-Zinsser disease - this disease could act as a source of new epidemics if louse infestation reoccurs (Bechach, Capo, Mege, & Raoult, 2008).

Impacts of Epidemic Typhus

In his book, Rats, Lice and History, Zinsser (1935) writes that epidemic typhus has caused more deaths than all the wars in history. Indeed, the widespread impact of epidemic typhus was demonstrated after World War I, where 20-30 million people died in Eastern Europe from epidemic typhus. During and after World War II, an additional several million died from this disease (Gross, 1996). These deaths were primarily located in Central and Eastern Europe, Northern Africa, Southern Italy, in which mass outbreaks occurred in concentration camps (Weiss, 1988). These concentration camps were unhygienic, overcrowded living spaces, which increased the likelihood of human louse transmission and infection. In China, 1 269 deaths were recorded in 1949. In the 1950s and 1960s, epidemic typhus cases and outbreaks involving body lice occurred frequently in poverty-stricken mountainous regions (Ming-yuan, Walker, Shu-rong & Qing-huai, 1987). Once again, these historical outbreaks occurred mostly in times of war or disasters in which prisoners, troops, soldiers often found themselves living in crowded conditions where they were unable to bathe or change clothes on a regular basis (Conlon). This disease is also more prevalent during colder months, where the cold weather may hinder people from regular bathing.

In present times, epidemic typhus cases are mostly recorded from Central and South America, as well as from rural regions of Africa, in which many live in poor sanitary conditions (Bechach, Capo, Mege, & Raoult, 2008). A large outbreak of epidemic typhus was recorded in Burundi in 1997, in which infection numbers were as high as 100 000 (Raoul et al, 1998). Moreover, scattered cases of epidemic typhus were also reported in Guatemala (Romero, Zeissig, España, & Rizzo, 1977).

Therefore, epidemic typhus is a disease that holds great public health significance, as it continues to affect communities in the world, especially in rural regions where living conditions may be crowded and unhygienic. This importance is illustrative of the continuing impact that lice has on society, particularly in their role in the transmission of diseases such as epidemic typhus in various populations in the world.

Trench fever

Trench fever is another disease which is transmitted through body lice, and outbreaks of trench fever further demonstrate the significant impacts of lice on society. This disease impacted millions of troops during war time, and similar to epidemic typhus, trench fever also continues to impact disadvantaged communities in the present day.

Trench fever is a disease that is characterized by headache, pain in the shins and/or loins and knees, chills, sweating, and dizziness (Byam et al, 1919). Other symptoms may include weakness, conjunctival injection, and the appearances of rashes (Bush, Vazquez-Pertejo, 2020). Similar to epidemic typhus, these symptoms can also range from being asymptomatic to fatal; this poses obstacles in the diagnosis of this disease, as symptoms vary considerably in their presentation.

Transmission of Trench Fever

The bacteria Bartonella quintana is the causative agent of the transmission of trench fever (Boutellis, Abi-Rached, & Raoult, 2014). Similar to the transmission mechanisms of epidemic typhus, the bacteria proliferates inside the bodies of human louse, and is transmitted into the human body through its presence in louse feces. The bacteria enters human skin primarily through wounds, and through scratching.

Impacts on Society

The impacts of lice and of trench fever are most notable in populations affected by war, such as civilians and soldiers during World War I. During the Great War, more than one million troops were affected by trench fever (Anstead, 2016). The disease was actually first named in World War I, where the disease affected German and Allied troops who found themselves crowded in trenches (Badiaga & Brouqui, 2012). In the spring of 1915, British and German medical officers, in their respective armies, noticed a five day relapsing fever in many of their patients. Many of the sick patients complained of a persistent pain in their shins. Moreover, it was observed that this condition was only present in British troops in the trenches, and not troops in the rear, and hence the disease was named trench fever (Swift, 1920). This can be explained by modern scientific analysis, which reveals that bacteria and body lice are most easily transmitted when there is persistent close contact with others in a crowded, unsanitary living situation; these living conditions are characteristic of the environment in trenches. It is hypothesized that the disease was imported from the Eastern front by German soldiers in 1914 (Maurin & Raoult, 1996) .

It is evident that trench fever had a great impact during World War I, as the illness affected the manpower resources of the Allies and the Central Powers (Byam & Lloyd, 1920). This is because in addition to the many deaths that resulted from this disease, soldiers affected by trench fever were deemed unfit for duty for 60 - 70 days (Anstead, 2016). It is estimated that in the British Army, trench fever caused a fifth to a third of all illnesses; it is also estimated that in the armies of the Central Powers, trench fever was responsible for a fifth of all the illnesses in their soldiers (Strong et al, 1918). After World War I, the impact of lice on society subsided, as the number of cases of trench fever decreased substantially. During World War II, the disease reemerged as soldiers once again found themselves in poor, crowded living conditions in the trenches.

In the early 1990s, the urban homeless populations of developed countries were also a particular target for trench fever, as people lived in poor and unhygienic conditions which was a catalyst for body lice transmission (Brouqui and Raoult, 2006).

Nowadays, the number of cases of trench fever has greatly diminished compared to the number of cases in World War I and World War II. However, infections linked to the bacteria Bartonella quintana and the human louse continue to affect socially disadvantaged persons, particularly the urban homeless populations. Ohl & Spach (2000) state that this is due to the crowded and unsanitary living conditions, as these environments promote close contact and exposure to individuals who may be carrying the bacterium Bartonella quintana. Therefore, although the impact of lice and of trench fever no longer affects the majority of communities, it is still crucial to place medical significance and improve public health efforts to target these disadvantaged communities, and to prevent infections in such populations.

Relapsing Fever

Lastly, relapsing fever is another disease that involves transmission through body lice, and further exemplifies the significant impacts that lice, particularly body lice, can have on society. The main symptom of relapsing fever is severe, recurring fever, and in some patients, neurological symptoms may also be observed (Larsson, Andersson,Bergström, 2009). Other symptoms also include coma, stiff neck, falling body temperature, and low blood pressure (U.S. National Library of Medicine, 2021). These symptoms may range in their severity, from asymptomatic to potentially fatal. It was also found that these symptoms are often most severe in younger children (Southern, Sanford, 1969).

Transmission

The link between human body louse and relapsing fever was first discovered and reported in 1907 by MacKie (Bryceson et al, 1970). Louse-borne relapsing fever is caused by the bacteria Borrelia recurrentis, which is transmitted by body louse (Boutellis, Abi-Rached, & Raoult, 2014). In particular, the bacteria is transmitted to individuals by its presence in body louse feces, which enters the human body through abraded skin and scratching

(Larsson, Andersson, Bergström, 2009). The transmission of relapsing fever is more likely in conditions such as war, famine, natural disasters, malnutrition, overcrowding, and poor general health (Southern, Sanford, 1969).

Impacts on Society

Much like previous diseases, relapsing fever most significantly impacts civilians as well as military populations which are disrupted by war and disaster (Boutellis, Abi-Rached, & Raoult, 2014). During World War I, relapsing fever impacted half a million people in Serbia. Between the years of 1919 and 1923, 13 million cases of the disease were reported in Russia and Eastern Europe during their civil war (Bryceson et al., 1970). A paper by Raoult & Roux (1999) expands on the severe impacts of relapsing fever during war time. In West Africa, thousands of cases were reported during World War I and World War II, and this led to many deaths in the area. In North America, one million cases were reported, most of which were from Algeria, Tunisia, Morocco, and Libya; the fatality rate in these regions due to relapsing fever was estimated to be 10%.

Relapsing fever is also prevalent in areas where people live in close proximity to ticks or lice habitats; for instance, relapsing fever is a major disease in sub-Saharan Africa (Larsson, Andersson, Bergström, 2009). A study conducted by Nordstrand et al (2007) found that in Togo, 8.8% of patients with fever were found to have relapsing fever. In these regions, an additional problem lies in the lack of available diagnostic tools in rural hospitals, which often leads to misdiagnosis and mistreatment of the disease (Larsson, Andersson, Bergström, 2009).

Therefore, it is made clear that the effect of relapsing fever is still a significant health care problem to this day. For instance, travellers can acquire relapsing fever and bring the disease to regions where the disease is not epidemic. Such travellers may have been exposed to the disease and may have been infected by the bacteria Borrelia recurrentis through activities such as camping, caving, or military training (Dworkin, Schwan, & Anderson, 2002). Thus, given the risk of relapsing fever recurrence in today's society, it is once again made clear that lice continue to have macroscale impacts on society, particularly in terms of their ability to transmit diseases such as relapsing fever. It is therefore important to inform populations at risk for these diseases on prevention methods - these methods mostly consist of maintaining good hygiene, and avoid louse infestation by washing clothing and hair in a frequent manner.

Chapter Summary

Throughout this chapter, it has been made evident the large scale impacts that body lice can have on society, primarily because of the ability of body lice to become vectors for bacteria that can cause life threatening diseases. This is because pathogens such as Rickettsia prowazeki, Borrelia recurrentis, and Bartonella quintana are able to be transmitted to humans through body lice and through the infestation of wounds. Diseases of focus included epidemic typhus, trench fever, and relapsing fever. All of these diseases caused mass deaths throughout times of war and disaster, particularly during times of war. This led to significant impacts on war efforts as well as social impacts on the affected regions and armies. These diseases also are prevalent in the present times, particularly in regions where poor, unhygienic living conditions are common. This is because body lice transmission, and in turn the transmission of the aforementioned diseases, increase in likelihood in conditions of poverty, malnutrition, and overcrowded living areas. These regions include the rural regions of Africa, where many hospitals lack the necessary diagnostic tools to prevent the misdiagnosis and mistreatment of these diseases. Given the severe and continuing impact of body lice on society, it is crucial for health efforts to concentrate on eradicating such cases to prevent mass infections, and to promote better living conditions in areas where such body lice proliferate.

Why is it important to study lice?

Diana Amiscaray

Introduction

Upon first impression, one might think that lice are simply pests that cause an itchy scalp. One might also believe that having lice is completely harmless and avoidable with proper hygiene. By the end of this book, it will be made evident how this impression is partially true and partially false.

The key to communicating accurate diagnoses, risks and effective treatments for lice, can be to better understand lice and their behaviour. While it is common knowledge that lice affect humans, especially children in their early years of elementary school (Paediatrics & child health, 2004), there are over 550 species of lice, each of which are host-specific (Bonilla et al, 2013). Of the 550 species of lice, only three species affect humans: Pediculus humanus capitis, Pediculus humanus humanus, and Pthirus pubis. While head lice can be considered irritating yet harmless, body lice have the potential to spread pathogens that invoke severe diseases including trench fever, relapsing fever, epidemic typhus and the plague. As mentioned in chapter three, body lice can be a vector for Bartonella quintana, Borrelia recurrentis, Rickettsia prowazekii and Yersinia pestis, all of which are deadly pathogens.

It is also important to note that humans are not the only ones targeted by these ectoparasites, as there are even species of lice that infest fish and thus pose a threat to the fishing industry. Infestation of fish by lice can cause a grave impact on sea life, consumers and the economy. Thus, the behaviour and transmission of lice is an important consideration to make when working in the fishing industry.

Further, because humans are not the only organisms affected by lice, studying the phylogeny of the lice that reside on other organisms can hint at their evolutionary history. Observing lice infestations may even outline behaviours of earlier groups, and prove handy in autopsy investigations.

This chapter will discuss the importance of studying lice as a vector for disease, and what factors should be considered when developing a strategy for diagnosis and treatment. As well, the chapter will discuss the effects of lice infestation on sea life, and how lice can be used as markers that unravel mysteries about human migration, behaviour and evolution.

Terminology

When discussing these organisms, it is common to refer to them as lice. Lice is the plural form of the organism while louse is the singular form.

The terms Pediculus and pediculosis appear very similar, however, Pediculus is the name of the genus in which human head and body lice are grouped in, while pediculosis is the term for the infestation of lice. Pediculus humanus capitis is the scientific name for the species that resides on the human scalp. On the other hand, Pediculus humanus humanus is the scientific name for the species that resides on the human body. Body lice may also appear under the alias Pediculus humanus corporis in some literature. The third type of lice that affects humans does not belong to the same genus as head and body lice, and is called Pthirus pubis. Pthirus pubis can also be called by its common name, "crab louse," and it resides in pubic hair.

Lice are ectoparasite insects; parasites are creatures that rely on a host for nutrients, and the prefix "ecto-" specifies that the parasite is living on the exterior of the host's body. The infestation is beneficial for the parasite, who is dependent on the other organism to survive. However, the infestation is often at the detriment of the host, who does not receive any sort of aid or benefit from the parasite stealing its nutrients. Further on in the chapter, it will be illustrated that the severity of symptoms and complications caused by lice varies among hosts.

Life Cycle

This section will discuss the life cycle and behaviour of head lice, Pediculus humanus capitis.

Head lice depend on humans not just for food, but also for shelter, warmth and moisture (Pediatrics and Child Health, 2004). Once a louse is able to gain access to a host, it sustains itself by feeding four to eight times a day. During feeding, the louse sucks blood from the scalp of its host (Castelletti & Barbarossa, 2020). At the same time, the louse also injects its own saliva into the scalp, which causes the itchiness characteristic of lice infestation.

This itchiness can cause significant discomfort for some people, while others may not experience itching at all. If the lice are able to mate, the female louse can breed up to 6 eggs per day over a time period of a month (Castelletti & Barbarossa, 2020). These eggs would then hatch into nymphs around six to eleven days following oviposition. From there, the nymphs undergo two separate stages of molting to finally develop into mature, adult lice. The two molting stages typically occur within 8 to 10 days (Castelletti & Barbarossa, 2020). Although the lice are considered adults after two molts, differentiation into male or female lice is only established after a third molt. Under the lice's ideal conditions, where they are able to feed and are sheltered by their host, adult lice can live for a month. However, the lice life expectancy is drastically shortened to one or two days if they are not able to infest a human's head (Castelletti & Barbarossa, 2020). Thus, lice are truly dependent on their host, and the ability to find and reside on a host means life or death for these parasites.

Symptoms

It is possible for an individual infested with lice to appear asymptomatic, however, the common symptom of pediculosis is itching. Itchiness is the hallmark symptom of pediculosis and is caused by an allergic reaction to the saliva of the lice. Sensitivity to the components in lice saliva can be developed four to six weeks after the start of pediculosis (Castelletti & Barbarossa, 2020). Aside from this, head lice are considered benign and easily treatable after diagnosis. For more details about treatment and prevention, please refer to chapter six.

Diagnosis

Despite lice being ectoparasites that are visible to the naked eye, diagnosing pediculosis remains a challenge. It has been found that healthcare providers often misdiagnose suspected cases such that the number of reported infestations is underestimated or overestimated (Paediatrics and Health, 2004). This inaccuracy in reported numbers is due to the inability to distinguish whether or not the infestation is inactive or active. To declare a case as an active infestation, there must be at least one live louse found on the scalp. Otherwise, the infestation should be reported as inactive. An additional obstacle to the examination process is that lice are capable of rapid movement. In fact, lice can crawl at a speed of 23 cm per minute (Paediatrics and Health, 2004.) To increase the efficiency and accuracy of diagnoses, it is imperative to train healthcare providers to a level of expertise. In addition, microscopy may be a useful tool for the examination (Paediatrics and Health, 2004).

Vectors of Disease

As mentioned in chapter three, lice can serve as a vector for pathogens such as Rickettsia prowazekii, Borrelia recurrentis and Bartonella quintana. In addition, it has been hypothesized that lice have been involved in the transmission of Yersinia pestis. These pathogens cause severe diseases including trench fever (Bartonella quintana), epidemic typhus (Rickettsia prowazekii), relapsing fever (Borrelia recurrentis) and the plague (Yersinia pestis) (Amanzougaghene et al, 2020). Luckily, it has been established that head lice do not pose a threat to public health, as only body lice are known to be vectors of disease.

Studies

The effect of lice on aquatic life

As previously noted, there are more than 550 species of lice (Bonilla DL, Durden LA, Eremeeva ME and Dasch GA, 2013), with only three species affecting humans. This section of the chapter will discuss a particular species of lice that has been proven particularly problematic for the fishing industry.

The louse Lepeophtheirus salmonis is a species that feeds on salmonids residing in the marine environment (Thorstad et al, 2015). These lice pose a grave threat to salmon as they feed on their skin and muscle, as well as cause dysfunction of their osmoregulatory processes (Thorstad et al, 2015). Additionally, the infestation of these lice can cause anaemia, which is a condition characterized by a decreased level of red blood cells or hemoglobin. The bites on the flesh of the fish also put the salmon at greater risk for secondary infections. As a consequence of these complications, these fish are particularly vulnerable to lice infestations and may die as a result (Thorstad et al, 2015).

These Lepeophtheirus salmonis lice are naturally found in both the North Atlantic and North Pacific Oceans, however, those found in the Atlantic and those found in the Pacific are classified as two separate groups of subspecies. Salmon farming increases the amount of salmon lice; farms that are equipped with open net cages enable lice to gain access to their preferred host, and thus, can increase the production of infective salmon lice larvae (Thorstad et al, 2015). Outbreaks of infestation affecting salmon were initially observed during the 1960s in Norwegian Atlantic salmon farms (Thorstad et al, 2015). These reported infestations were noted to have occurred shortly after the use of cages in salmon farms. From then on, outbreaks of similar nature were reported during the mid-1970s in Scottish Atlantic salmon farms, and in Ireland between the years 1989 and 1991 (Thorstad et al, 2015).

Studying lice in salmon has thus prompted discussion of more effective practices for fish farming. By ensuring effective practices for fish farming, the industry may protect its sea life, consumers and avoid a disadvantageous effect on the economy.

Studying lice as a trace of evolutionary milestones

As emphasized in chapters one and two, the existence of lice is not a new discovery, and infestations have been documented since ancient times. With this knowledge, some researchers seeking to understand the migratory patterns and evolution of humans may opt to study lice infestations. The use of clothing can be regarded as a modern behaviour developed to adapt to colder climates (Toups et al, 2011). Provided that body lice can reside in the clothes of its human host, determining when body lice evolved from their head lice ancestors may provide insight to this evolutionary puzzle (Toups et al, 2011).

Another reason why lice are effective markers for evolution is due to their host-specific nature. Because a certain species of lice is limited to a specific host, they are demonstrated to evolve together, which provides researchers with an alternate tracking method that complements fossil records (Amanzougaghene et al., 2020). One of the most notable findings of studying lice phylogenetic trees is that human and chimpanzee lice share a common ancestor (Amanzougaghene et al., 2020). This finding is consistent with the theory that their respective hosts are evolutionarily related.

Studying Lice to Investigate Neglect and Cause of Death

While head lice are typically associated with poor hygiene habits, the infestation of head lice actually do not prove an individual has been living in unsanitary conditions. Despite this, forensic scientists have been able to prove details such as time and cause of death, and confirmed the patient's experience of neglect (Lambiase & Perotti, 2019). The presence of nit clusters in the patient's hair is an indicator of length and frequency of neglect (Lambiase & Perotti, 2019). In one case, results of the autopsy outline a severe infestation in an elderly woman sent to an emergency ward. Unfortunately, the woman passed away shortly after being admitted. The forensic team noticed how long the woman's hair was and decided to collect insect samples from her hair. The collected lice were all dead at the time of analysis, and it was observed that every piece of hair contained nits (Lambiase & Perotti, 2019). A cause of death was deduced under the knowledge that lice feed off of their host to survive. The woman was discovered to have been consuming nifedipine, an antihypertensive drug, on a daily basis, which notably did not interfere with the ability of the lice to reproduce. For a time period of two years, the lice have been able to thrive off of their host regardless of the nifedipine consumption. However, two months prior to her death at the hospital, the lice have not been able to reproduce at their regular rate, and they too were found deceased. The cause of death was then deemed to be excessive use of nifedipine (Lambiase & Perotti, 2019).

Chapter Summary

Lice are often regarded as harmless yet irritating insects that prey on humans, especially young children in primary schools. What the majority of people don't know is that humans aren't the only ones affected by lice. Rather, there are numerous types of lice that target other animals in order to survive. As ectoparasites, lice infest their preferred host, live on the exterior of their body, and suck their blood to obtain nutrients. While parasites thrive off of the shelter of their host, the infestation comes at the expense and detriment of the host. To humans, head lice can simply cause discomfort and itchiness. In contrast, salmon are particularly vulnerable to lice and can die of secondary infections, physiological stress and a variety of other complications.

Lice can also be a vector of disease, as human body lice can carry pathogens such as Rickettsia prowazekii, Bartonella quintana, Borrelia recurrentis and possibly Yersinia pestis. Studying lice is therefore crucial in order to maintain public health and safety and prevent the occurence of epidemics.

Aside from the clinical significance of lice and their role in disease, studying lice may provide insight on the evolution of organisms and human migration. Due to the fact that lice are host-specific, studying the relationships between different species of lice may suggest relationships between their respective hosts. This was exactly the case with human lice and chimpanzee lice, where the two species of lice shared a common ancestor, thereby suggesting that their hosts share an ancestor as well. Studying the evolution and genetics of lice has also revealed details about human behaviour over history, as some researchers have tracked the divergence of body lice from head lice as an indication of the date when humans began to wear clothing. The date of commencement of clothing use should in turn suggest the time period of migration from hot to cool climates.

Lice has also been useful as a tool for forensic scientists. Analysis for the presence of nits in the hair of the deceased patient suggested how long and how often she had been neglected. Additionally, analysis of the reproduction of lice residing in her scalp suggest that the cause of death was excessive use of nifedipine, an antihypertensive drug the woman was consuming on a daily basis.

Overall, it is important to study lice to be informed about the truth behind misconceptions about pediculosis, prevent potential disease outbreaks caused by the pathogens that body lice carry, develop sustainable practices in our fishing industries, and solve mysteries such as evolution and cases of neglect.

What are lice?

Ellen Mak

Overview

Lice, or plural for louse, are recognized as one of the most successful insects on the planet due to the parasitic relationships they establish with numerous warm-blooded hosts (Mullen and Durden, 2019). They infest mammals and wild birds as parasites or scavengers and are commonly found to affect the health of humans, pets, and livestock by feeding on host blood. While lice infestations are not dangerous, human-specific lice have indirectly influenced human history due to their ability to transmit pathogens, causing major epidemics like typhus, trench fever, and louse-borne relapsing fever as detailed in chapter 3. Despite their impact, most of around 5000 species of ectoparasitic lice have little to no medical or veterinary importance.

Using their scientific name, the Phthiraptera are obligate ectoparasites, meaning that they live on or in the skin of the host's body and must exploit their host to complete its life cycle and reproduce (Mullen and Durden, 2019). Among all known ectoparasites however, lice are unique because they are ectoparasitic throughout both juvenile and adult stages (Royal Entomological Society, n.d.). They remain on the host's body for their entire life cycle and are rarely found on their own. As a result, they have been observed to have a relatively profound host specificity, where one species of louse only infects one type or species of mammalian or avian (bird) host (Mullen and Durden, 2019). This degree of specificity may vary. For example, Trichodectes canis is a dog biting louse that attaches to the base of the individual hairs of domesticated dogs or wild canids, while Pediculus humanus capitis is the only head louse that dwells on human scalp (Mullen and Durden, 2019).

Taxonomy

Lice are scientifically named according to the order Phthiraptera, which comes after their kingdom Animalia, their phylum Arthropoda, their class Insecta, and their superorder Psocodea (Mullen and Durden, 2019). Traditionally, lice were classified based on two distinct morphological groups: sucking lice were placed within the order Anoplura and chewing or biting lice were placed within the order Mallophaga (Palma and Barker, 1996). Sucking lice pertains to those that feed on the blood of mammals, while chewing lice feed on the feathers, fur, and/or blood of both mammals and birds. However, phylogenetic analyses have enabled the combination of Anoplura and Mallophaga into the order Phthiraptera, which is now divided into four suborders: the Anoplura, Amblycera, Ischnocera, and Rhynchopthirina. Here, the Anoplura remains unchanged, while the remaining three suborders were extracted from the former order Mallophaga. This was done after the realization that chewing lice do not represent a monophyletic group; some members were found to be more closely related to sucking lice from the order Anoplura than to other chewing lice (Mullen and Durden, 2019). A monophyletic group refers to organisms within the same taxon that share a most recent common ancestor. According to Durden and Musser (1994a), approximately 550 species of sucking lice have been discovered, while Price et al. (2003a) reports 4464 species of chewing lice, 12.4% of which parasitize mammals. However, these numbers are not exact. It is predicted that there are many more undescribed species of lice, though the growing number of threatened and endangered avian and mammalian species also makes record keeping difficult (Galloway, 2019).

Sucking and chewing lice originated from a nonparasitic ancestral group within the superorder Psocodea (Mullen and Durden, 2019). These nonparasitic book lice and bark lice diverged 100–150 million years ago into chewing lice within the order Phthiraptera. Sucking lice further diverged 77 million years ago from chewing lice and later proliferated to form the many species that parasitize multiple orders and families of mammals and birds. As of now, the fossil record for the Phthiraptera only consists of one well-documented specimen. Found in Germany, the fossil was classified as an amblyceran menoponid chewing louse and dates to approximately 44.3 million years ago (Dalgleish et al., 2006). For reference, amblycera is the order, while menoponidae is the family that the louse is classified under.

Sucking lice of medical importance within the order Phthiraptera are separated into two families, the Pediculae, which comprise head and body or clothing louse, and the Pthiradae, which represent the crab or pubic louse (Mullen and Durden, 2019). Both families make up a total of three species of louse that feed on human blood, where the head, body, and crab louse are named Pediculus humanus capitis, Pediculus humanus humanus, and Pthirus pubis, respectively. Sucking lice of veterinary importance are established under five families: the Haematopinidae, Hoplopleuridae, Linognathidae, Pedicinidae, and Polyplacidae. As for chewing lice, those of veterinary importance are the Boopiidae, Gyropidae, Menoponidae, Philopteridae, and Trichodectidae. Since lice are highly host-specific, each species within a family affects only one host type or species. In terms of those with veterinary significance, this includes pets and livestock like guinea pigs, cats, dogs, domestic fowl, cattle, horses, sheep, and many more.

Morphology and Diversity

As stated previously, lice are divided into two morphological groups: sucking lice that feed on the sebaceous secretions and bodily fluids of their host, while chewing or biting lice are scavengers that feed on the skin, feathers, hair, fur, and other debris found on the body of their host (Mullen and Durden, 2019). Since lice are obligate ectoparasites that cannot live on their own, they have developed high host-specificity and have gradually co-evolved with them. Some are so specific that they infest only a specific part of the host's body. Interestingly, some animals are also known to host multiple species of lice, where mammals are typically seen with one to three, while birds are seen with two to six. In some cases, the numbers can reach up to fifteen species.

According to the Greek translation, Phthiraptera comprises the words 'phtheir' and 'pteron', where the former means louse and the latter means wing (Mullen and Durden, 2019). Although lice have winged ancestors, lice are in fact wingless insects that range between 0.35–11 mm long as adults. Their bodies are dorsoventrally flattened with abdomens that possess variably sclerotized plates which provide the rigidity needed for distension, or swelling of the abdomen, after their meal. Lice abdomens are also adorned with numerous setae, which are stout hairs or bristles that help anchor the insect. In female lice, their abdomens terminate into genitalia that are specialized with two pairs of gonopods. These serve to guide and glue eggs, which are called nits, onto the hair or feathers of the host using specialized saliva. These eggs only hatch when the body temperature of the host is sufficiently high (Royal Entomological Society, n.d.). In male lice, their genitalia are highly variable, but are observed to have a relatively large and sclerotized pseudopenis as well as two to four testes (Mullen and Durden, 2019). After mating, females ensure that the egg is heavily chitinized to protect the embryo from mechanical damage and desiccation. In terms of immature lice, their bodies are a smaller version of adult lice but lack genitalia.

Furthermore, adult lice have reduced or absent eyes and short legs bearing well-developed claws for attaching to their host (Royal Entomological Society, n.d.). Lice with mammalian hosts bear one claw, while those that parasitize avian hosts bear two (Smith, 2011). Their antennae are short, small, and segmented into three to five parts, though in rare cases they may be absent. Within the four suborders, sucking and chewing lice have very distinct mouthparts (Royal Entomological Society, n.d.). Generally, chewing lice have mandibulate mouthparts used for biting or chewing the skin, hair, fur, or feathers of their host. This produces irritation in the host that causes them to scratch the site, leading to small wounds that the lice can later feed on. In amblyceran and ischnoceran chewing lice, their heads are much larger than their thorax or chest region, which is the opposite of anopluran and rhynchopthirinan chewing lice (Mullen and Durden, 2019). Some species of chewing lice also house symbiotic bacteria within bacteriocytes that are necessary for digestion. Analysis of the four suborders of chewing lice show that while they all possess chewing mouthparts, each differ in their specialization and mechanics. For instance, rhynchopthirinan chewing lice are special because they possess a tiny mouth at the end of their elongated head. Other modifications within the suborders include species of chewing lice with mouthparts that also function as sucking organs. Moreover, sucking lice have mouthparts that are highly modified for sucking or piercing (Smith, 2011). These hematophagous, or blood-feeding insects have heads that are much smaller and narrower than their thorax. In the more specialized anopluran sucking lice, their mouthparts contain recurved teeth that sink into the mammalian host skin during feeding (Mullen and Durden, 2019). Once the host blood vessels are punctured, the sucking lice injects anticoagulants and other enzymes via their saliva into the host to allow them to feed on the blood directly. According to research on the well-known human body louse, the internal anatomy of sucking lice was found to be like most hematophagous insects. This includes similarities like strong cibarial, which pertains to the preoral cavity, and esophageal muscles.

As an obligate parasite, lice have evolved various adaptations and strategies to exploit their hosts to survive (Mullen and Durden, 2019). To remain in close contact with their host, lice are very tiny in size, possess claws that cling to host skin, hair, fur, or feathers, and have a flattened body. Particularly in hosts involved in grooming activities, are arboreal, or can fly, their claws tend to be more robust to facilitate host attachment. Lice have also evolved to cryptically match the colouration of the pelage (fur or hair) and plumage (feathers) of their host (Smith, 2011).

Life History and Distribution

Lice are hemimetabolous insects that develop through incomplete metamorphosis from larva to adult but have no pupal stage (Mullen and Durden, 2019). This includes three distinct stages: the egg, nymph, and adult stage. After the egg hatches, there are three nymphal instars, or developmental stages, which are separated by insect moults that occur when they shed their exoskeleton and grow. Although this life cycle varies between species, the egg stage typically lasts 4–15 days, each nymphal instar lasts 3–8 days, and the adult phase may reach up to 35 days. Lice can optimally achieve 10–12 generations, though factors like host grooming, immune responses, feather loss, hibernation, predators, parasites, and unfavourable weather conditions can greatly reduce this number. The fecundity, which refers to the ability to produce abundant offspring, of female lice ranges from 0.2 to 10 eggs per day. While most lice reproduce sexually, some species are known to reproduce through parthenogenesis, or asexually without fertilization from sperm. An example of this are lice from the genus Damalinia which cause hair loss syndrome in North American deer and constitute a small population of cattle biting and horse biting louse.

Lice distribution typically mirrors that of their hosts, since they cannot survive without them (Smith, 2011). Other factors that influence distribution include the fit of the host and the host specificity of lice. Young and unfit hosts tend to house larger populations of lice, while those that develop immunity are seen with lower populations (Mullen and Durden, 2019). Consequently, host diversity is a better predictor for louse diversity and distribution than geographical or ecological correlates (Smith, 2011). However, some anomalies are present that affect lice distribution. Although almost all bird species have lice, some mammals do not, including monotremes, pangolins, bats, and whales (Royal Entomological Society, n.d.). Some theories suggest that lice may be sensitive to the temperature changes that come with bat hibernation, and that lice would not inhabit marine mammals due to their immersion in sea water. Though this may be a possible explanation, the fact is that lice do occur on some hibernating mammals, and marine mammals like seals do inhabit a species of lice that breed when the seals come on land.

Behavior and Ecology

In addition to being morphologically distinct, the feeding behaviours of chewing lice and sucking lice are also one of a kind. Chewing lice feed by biting or scraping the hair, fur, and feathers with their mandibles before forcing the small pieces into their mouths (Mullen and Durden, 2019). They can also feed on integumental skin debris and secretions, or in some cases can imbibe blood by scraping on host skin. The rhynchophthirinan Haematomyzus elephantis is a facultative hematophage that feeds on the blood of both African and Asian elephants, making it an intermediate of both chewing and sucking lice. Sucking lice on the other hand must imbibe blood to ensure its development and survival. They do this by contracting blood through a piercing hollow stylet on the hypopharynx using cibarial and pharyngeal muscles. Sucking lice also house symbiotic bacteria that synthesize essential vitamins for survival. Depriving a female louse from these vitamins can cause it to become sterile, since some symbionts migrate to the ovaries.

As lice are highly dependent on their host, certain conditions or types of lice severely affect their long-term survival (Mullen and Durden, 2019). Deprivation of blood for even a few hours can be fatal for sucking lice, though chewing lice can survive for several days away from the hair, fur, or feathers of their host. Chewing lice that are also hematophages also cannot survive for prolonged periods away from their host. Furthermore, survivability depends on environmental conditions, where off-host survival is greater at low temperatures and high humidities. Interestingly, amblycerans are more likely to be encountered away from their host than any other suborder due to the observation that they are found crawling on the nests and eggs of birds.

Lice exhibit varying degrees of host specificity. Some species only parasitize one host species, like the hog louse and large turkey louse, while some species can parasitize two or more closely related hosts, like the horse sucking louse that infests horses, donkeys, mules, and occasionally zebras (Mullen and Durden, 2019). Lice found on atypical hosts are commonly termed stragglers. In addition, species like the sheep foot louse and sheep face louse parasitize a certain area of the host's body. As a result, lice have evolved morphological adaptations to the specific conditions and attributes of the host site, which include differences in the pelage or plumage, thickness of the skin, grooming activities of the host, and the availability of host blood vessels. An example of this is seen in the high prevalence of site specificity in sedentary, specialized ischnocerans over mobile, unspecialized amblycerans. Some species also inhabit very specialized host sites, like the inside of oral pouches of pelicans or the feather quills of several species of birds.

The population changes of lice vary according to seasonal trends. Lice on small or medium-sized mammals have been observed to exhibit minor changes in population levels over the year, while lice on larger mammals show clearer seasonal trends in population (Mullen and Durden, 2019). These patterns are attributed to host molting, fur density and length, hormone levels in the blood, and climatological factors like high summer temperatures, sunlight, and desiccation. Most of the time, louse populations in the temperate regions increase during the winter months and decrease during the summer months. An exception to this is seen with the cattle tail louse that increases in population during the summer.

Lice transfer occurs between hosts primarily through direct host contact (Mullen and Durden, 2019). This happens when lice transfer from an infected mother to her offspring during suckling and is seen with sheep face louse and sheep biting louse. Lice transfer also occurs through physical contact during mating or fighting between hosts. However, direct contact is not always necessary; the sheep foot louse can transfer from its original host to a new host after several days on pastureland. Another mode of transmission happens through phoresy, where lice temporarily attach to arthropods from one host to another. It is more common for lice to attach to mobile, hematophagous arthropods like flies, and is seen more frequently on chewing lice over sucking lice since attachment is less efficient with the robust claws of sucking lice.

Mating between male and female lice occurs on the host and is initiated when the male louse pushes his body beneath that of the female and curls the tip of its abdomen upward (Mullen and Durden, 2019). While most lice exhibit similar mating behaviours, this varies for certain species. For example, the human crab or pubic louse continues to cling onto the host hair rather than to each partner during mating, and some male ischnoceran chewing lice grasp the female louse using hooklike antennae during copulation. Once the eggs are ready to be laid, female oviposition behaviour involves crawling to the base of hairs, furs, or feathers and cementing one egg at a time close to the surface of the skin. The finger-like gonopods of the female genitalia guide this process and glue the egg using a specialized substance. Since louse embryos are temperature-sensitive, the female louse must lay their eggs in the area that meets this requirement.

Chapter Summary

Currently, there are about 5000 species of louse from 24 families worldwide (Royal Entomological Society, n.d.). The Phthiraptera comprise of unique ectoparasitic insects that spend their entire life cycles on the outside of their hosts. As a result, research on lice becomes difficult as it can only be done by studying their mammalian and avian hosts, though much insight has been gained on the coevolution of lice and their relevant hosts. While lice tend to be negatively viewed due to their medical and veterinary implications on humans, pets, and livestock, they are quite important for research and have a long-standing phylogenetic history.

What are common types of lice today? What are treatments used to treat them?

Jessica Henry

Introduction

Lice can be divided into two categories, chewing lice and sucking lice. Chewing lice encompasses the suborders of Amblycera and Ischonecara(Britannica, 2011). Chewing lice can be from 1mm-5mm in length, and can range in colour from white to black. Animals such as marsupials, rodents, and birds are affected by chewing lice, but humans are not(Britannica, 2011). They are called chewing lice because they feed on fur, dried blood, and skin debris and secretions(Britannica, 2011). The second group would be sucking lice, which has the suborder of Anoplura. Anoplura affects both humans and other animals. Sucking lice feed by piercing the skin of their host and sucking the blood from their blood vessels(Britannica, 2009). The three most common types of lice and the lice that affect humans are Pediculus humanus capitis(head lice), Pthirus pubis(pubic lice), and Pediculus humanus humanus(body lice).

Head Lice

The most common type of lice is head lice. Head lice are a part of the Anoplura suborder that encompasses sucking lice. Head lice need to constantly and consistently feed, meaning they have to stay on their hosts body. They only leave to transfer to another host, though sometimes they are removed purposely by humans, or accidentally through mechanical activity. If the louse is removed from the head and does not find a host within 1-3 days, it will die as both the lack of food and low temperatures negatively affect their survival (CDC, 2020)(MSU Pesticide Safety Education Program, 2006). The presence of head lice does not indicate poor hygiene, it is usually the result of a wider outbreak. Head lice cannot hop or jump, and they are wingless creatures so they cannot fly either, leaving crawling as their only mode of self transportation (MSU Pesticide Safety Education Program, 2006). Their transmission is based on people sharing items that touch their scalp like hats, brushes, or bedding, and they can also crawl from head to head during periods of close proximity such as hugs (Mayo Clinic Staff, 2020). The ones most susceptible to head lice infections are school aged children (primarily preschool and elementary) and those in close proximity to them (teachers and guardians)(Healthline Editorial Team, 2019). This is because they are more likely to play together, share items. Head lice infections are also more prevalent in women and people of Caucasian descent (Winnipeg Regional Health Authority, 2008). Women get head lice more often due to social behaviors such as sharing clothes and hair accessories (Guenther, 2021). Those of African descent are the least likely of all races to get head lice. This is because the claw of the head lice, which is the way they hold onto the hair, is more adapted to the round hair shafts that people of asian and caucasian descent have; which is why head lice is more common in people with caucasian ancestry (Guenther, 2021).

The vast majority of head lice is found on the head, though on rare occasions they can be located on the eyelashes and eyebrows (CDC, 2020). Head lice prefer warmth, so they are usually near the ears or the neckline (CDC, 2020). The warmer the area the more blood flow, which is the sole source of the head lices sustenance, so their affinity for hotter areas of the head makes sense. Head lice lay their eggs close 1-4mm away from the scalp so that they are close to both a heat and food source (Winnipeg Regional Health Authority, 2008). It takes 6-12 days to go from nit to nymph, 7-14 days to go from nymph to adult, and they can be adults for 7-30 days (Winnipeg Regional Health Authority, 2008)(CDC, 2019). The female head louse lays 3-10 nits per day, amounting to 100-300 nits she can possibly lay (Winnipeg Regional Health Authority, 2008). Nits are 0.3mm-0.8mm, usually oval shaped (though can be teardrop or flask shaped), and can be: grayish-white, early, silvery-white, yellow-white, transparent, glistening, or opalescent (Winnipeg Regional Health Authority, 2008). Only nits laid by inseminated females will hatch and of those 2-12% will not

(Winnipeg Regional Health Authority, 2008). The nit then develops into a nymph which is the intermediary stage of the head louse. The nymphs are like a smaller version of adult lice, measuring 1 mm long (Winnipeg Regional Health Authority, 2008). Before becoming adults they molt three times. The adult is the last stage of the louse life cycle, living for up to 30 days (CDC, 2019). They are dorsoventrally flattened, have four short antennae, anterior piercing mouthpieces to pierce the scalp and feed, and six stubby legs with claws on them used to attach to hair shafts (Winnipeg Regional Health Authority, 2008). They can be 2-4mm long, and this colour is determined by the colour of hair on the scalp that they are occupying, though they become red coloured after feeding (Winnipeg Regional Health Authority, 2008). Mating does ot occur until the lice are fully grown, and they do not leave the head to find new hosts until they are fully grown either (Winnipeg Regional Health Authority, 2008).

Since head lice are a part of the sucking lice family they feed through biting the skin barrier and sucking blood. The head lice inject both a local anesthetic and an anticoagulant while they feed (Winnipeg Regional Health Authority, 2008). This allows the host to not feel anything and also prevents the blood from clotting so they can properly feed. Head lice can feed on the blood multiple times a day, but the blood does make them bloated and they keep their original size (MSU Pesticide Safety Education Program, 2006). Head lice do not carry or cause any diseases (CDC, 2019). Infections can first be asymptomatic, but itching can occur caused by an allergic reaction to the anticoagulant secreted by the head lice when it is feeding (CDC, 2019)(Winnipeg Regional Health Authority, 2008). If the scratching causes open wounds secondary infections can occur. The symptoms for a head lice infection are tickling on the scalp, itching and subsequent sores, and difficulty sleeping (CDC, 2019).

Head lice can be diagnosed either by finding a live louse or nymph, or by the presence of viable nits. Head lice are photosensitive and move quickly, making them hard to see with the naked eye, they appear darker on people with darker and lighter on people with light hair, making them even more difficult to see (CDC, 2020). Nit combs and magnifying glasses are used to find adult louse and nymphs in the hair (CDC, 2020). If no louse or nymphs are found then the second option would be to look for viable nits. Viable nits are nits that have not hatched yet are located 1-4mm from the scalp (Winnipeg Regional Health Authority, 2008). They are also found using nit combs. Nits can be confused with other things such as dandruff or dry shampoo flakes, but those are easily movable unlike nits (Healthline Editorial Team, 2019). A woodlight can also be used to tell if a nit is really a nit. If shone on the hair a nit will turn blue, and viable nits can be identified using a microscope (Mayo Clinic Staff, 2020). Nits are very secure on the hair strand, so the length of the infestation can sometimes be told by how far away the nit is from the scalp (Winnipeg

Regional Health Authority, 2008). The average hair growth is 0.37 mm per day and viable nits are placed from 1-4 mm above the scalp, so by calculating the average movement is it possible to tell how long the infestation has been present for (Winnipeg Regional Health Authority, 2008).

There are many treatments for getting rid of head lice, falling into the categories of over the counter medications, prescription medications, and natural remedies(though they are not all proven to work). The two most used over the counter medications are Pyrethrin and Permethrin. Pyrethrin is a pesticide that is extracted from chrysanthemum flowers (Healthline Editorial Team, 2019). It is usually combined with additives that make the pesticide stronger, though side effects include itching and redness of the scalp (Healthline Editorial Team, 2019). Pyrethrin can only kill nymphs and adult lice, so it is recommended to use Pyrethrin 9-10 days after the first round to kill all of the nits that may have hatched (Mayo Clinic Staff, 2020)(CDC, 2019). Permethrin is a synthetic version of Pyrethrin which also does not kill nits (Mayo Clinic Staff, 2020). The same repetition and side effects as Pyrethrin are present with Permethrin use. Prescription drugs are used when drug resistant head lice are found. The most common prescription drugs are Benzyl alcohol lotion, Ivermectin lotion, and Malathion lotion; others include Spinosad topical suspension and Lindane shampoo (CDC, 2019). Benzyl alcohol lotion kills nymphs and adult lice but not nits which means it requires a second round a week after the first initial application. Ivermectin lotion is single use and is toxic to all stages of lice, there is also a tablet form of Ivermectin (Mayo Clinic Staff, 2020). Malathion lotion kills live lice and some nits, if lice is present after 7-9 days a second treatment is recommended (CDC, 2019). Non medical treatments include wet combing, which is when the hair is wet and lubricated and run through with a nit comb, and the use of smotherthing agents such as mayonnaise and petroleum jelly (Mayo Clinic Staff, 2020). Both give varying results and are not medically recommended treatments for head lice infections. A more consistent natural remedy would be the use of essential oils such as tea tree and anise oil, which kill head lice by suffocation (Mayo Clinic Staff, 2020).

Head lice can be prevented by avoiding close head-to-head contact with others and by not sharing materials that are in contact with your scalp such as hats, hair accessories or brushes. You can disinfect brushes by soaking them in water that is at least 130°F for 5-10 minutes and clothing by heating them to 130°F in the dryer (CDC, 2019). For non-washable clothes there is the option of dry cleaning or sealing the clothes in a bag depriving the head lice of food (CDC, 2019). Lastly, you can vacuum and clean up after known infected people, and also avoid close contact with them as they are curing their infection.

Pubic Lice

Pubic lice is considered a sexually transmitted infection (STI). They are often called crabs due to the shape of the lice, with a rounded body and enlarged front legs that look like crab pincers (CDC, 2020). Pubic lice is transmitted through contact of pubic hair (Crabs, 2020). There does not need to be any penetration or other sexual contact involved as the lice lives on the hair, not in sexual fluid or on the genitalia. It can also be transmitted by sharing items such as towels and bedding with someone who is infected. Pubic lice in children can be a sign of sexual abuse (Mayo Clinic Staff, 2020). Like head lice, pubic lice cannot jump or fly, and rely on crawling from place to place. You are most susceptible to getting pubic lice if you are sexually active, especially if you have more than one partner (Brazier, 2020). If a person has pubic lice they should also be screened for other STIs (CDC, 2020).

Pubic lice is mainly found on hair in the pubic region, though they have an affinity for coarse hair. This means that they are also found on eyelashes, eyebrows, beards, moustache, chest hair, and armpit hair (Crabs, 2020). Pubic lice lay their eggs near the base of the hair, as warmth and food are two aspects crucial to their survival, and the base of the hair provides access to both (Crabs, 2020). Pubic lice nits take 6–10 days to hatch, they are nymphs for 2–3 weeks, and like head lice can live for up to 30 days as an adult (CDC, 2020). The female pubic louse will lay about 30 eggs in her lifetime (CDC, 2019). Nits are oval shaped and range in colour from yellow to white (CDC, 2020). Like head lice, pubic lice nymphs look like smaller adult lice, and require three molts to become an adult. Adult pubic lice are from 1.5mm-2.0mm long, and are flattened and more broad than other types of lice (CDC, 2019). They resemble crabs with their pincers like front legs.

Pubic lice are sucking lice and feed exactly like head lice. Also like head lice they will die if they do not get to feed for 1–3 days, this usually occurs when a louse has lost its primary host and can not find a new one in time. Unlike head lice, pubic lice feed from the genital region, using its crab-like claws to grab onto the coarse hair in the area to anchor them, and then injecting their anesthetic and anti-coagulant to start feeding. Pubic lice does not itself carry any diseases, but it is sometimes accompanied with other more serious STIs, so it is recommended that infected persons should be screened (Mayo Clinic Staff, 2020). The louse however does cause itching due allergic reactions to their bite which may result in the development of eczema (Britannica, 2019). The most common symptom of pubic lice is gential itching, though this is a common symptom of most STIs.

To be properly diagnosed with pubic lice, a live louse or a viable nit usually needs to be found in the area (CDC, 2020). A magnifying glass or microscope can be used as an aid in the search for the lice, though pubic lice are sometimes large enough to see with the naked eye (Crabs, 2020)(CDC, 2019). Methods used to diagnose head lice can also be used for pubic lice.

The same over the counter drugs, Permethrin and Pyrethrin, are approved to be used for pibic lice. Malathion, Lindane shampoo, and Ivermectin are all prescription drugs that can be used to fight a pubic lice infection (CDC, 2019). For these treatments to work the area must be thoroughly washed and all clothing cleaned to get rid of any lice that are attached to it(CDC, 2019). Sexual partners should also be contacted and should also get treatment. If the infected person cannot get the lice removed on their own then an appointment with a doctor should be made.

Unlike other STIs condoms give minimal help in the prevention of pubic lice infection, this is because condoms do not protect the area that pubic lice inhabit (Brazier, 2020). To control the spread, infected persons should cease sexual contact and sharing potentially infected objects and clothing with others. Bed sheets and towels should be disinfected with clothing, this can happen by doing laundry and drying the articles at at least 130°F or dry cleaning the clothes (CDC, 2019). The only way to prevent pubic lice infection in the context of sexual contact is abstinence.

Body Lice

The last most common type of lice is body lice. Unlike head and pubic lice, body lice do not live on their hosts. Homeless people, refugees, and people displaced by natural disasters are most likely to get body lice infections (Mayo Clinic Staff, 2020). Body lice can only crawl, and spreads through close person-to-person contact and sharing clothing and bedding (Vafasso, 2019).

Body lice live on clothes and bedding and are found in places where hygiene is difficult to maintain, such as homeless shelters and refugee camps (Mayo Clinic Staff, 2020). Contrary to the habits of both head and pubic lice, body lice prefer to lay their nits in clothing seams and on bedding (Mayo Clinic Staff, 2020); though they rarely lay nits on body hair (CDC, 2020). Female body lice can lay up to eight nits per day (CDC, 2019). Body lice nits hatch after 1-2 weeks, and they take 9-12 days to mature into adults (CDC, 2020). The nits are oval, yellow to white, and usually 0.8mm by 0.3mm in size, the nymphs are like smaller adults and need to molt three times to become adult body lice (CDC, 2020). The adult body louse lives for up to 30 days, and has six legs and is tan to a grayish white (CDC, 2020).

Since body lice do not live on the human host, they travel multiple times a day to the host to feed on their blood (Mayo Clinic Staff, 2020). Like other sucking lice, they use anesthetics and an anticoagulant injection to help them feed undisturbed. Itching is a symptom of body lice, blood can be drawn and crust over due to the intensity of the itching (Mayo Clinic Staff, 2020). Rashes can appear due to the allergic reaction from the bite of the lice, red bumps, thickened skin, and hyperpigmentation caused by scars if the infection is present for long enough (Vafasso, 2019). The thickening of the skin and discolouration is called vagabonds disease, and usually occurs around the waist, groin, or upper thighs (CDC, 2020). Like other lice infections, body lice themselves do not usually have any known diseases, though other infections can get in through sores caused from scratching. Rarely, body lice can carry typhus or louse borne relapsing-fever, though this only happens in places where good hygiene is not possible (Vafasso, 2019). Body lice infections are diagnosed like other lice infections, through the presence of viable nits or lice moving lice. Body lice can also be diagnosed by finding lice on the seams of clothing and materials such as bedding (CDC, 2019).

Body lice does not usually need medication to remedy, and can be fixed through a short google search and a trip to the bathroom and laundry room or local laundromat. To get rid of body lice, infected people need to wash themselves and their clothing and bedding, and anything else that they think may be contaminated (Mayo Clinic Staff, 2020). Having the dryer at at least 130°F will get rid of the body lice. If lice is still present the use of permethrin is recommended, if that does not work then infected people are urged to see doctors (Mayo Clinic Staff, 2020).

To prevent and control the further spread of body lice, one can practice good hygiene by bathing daily and doing laundry once a week changing into clean clothes daily (CDC, 2019). Do not share clothing or bedding with others, as those are the usual places where body lice and their nits are found.

Conclusion

Human inflicting lice can be easily combatter by either removing it yourself, decontaminating the lice habitats, or through the use of medication, either over the counter or prescription. Head lice is the most common type of lice, and is the one that is usually referred to when lice outbreaks happen in schools. It is the easiest to transmit and maintain. Head hair is more exposed than human hair, giving head lice an advantage in transmission. Body lice is easily controlled by hygiene, whereas head lice are more resilient and can stay in the hair even while it's being washed.

Chapter Summary

All three types of lice are very similar. They are all sucking lice, meaning they feed off of the blood of their host. They can be exterminated with a plethora of medications, and also using heats higher than 130°F. Human lice does not usually carry diseases or infections, but people that contract lice are susceptible to infections due to the open wounds that can be caused from excessive scratching. Lice can only be surely diagnosed by finding live lice or viable lice nits that have not hatched. Lice can only crawl, limiting their motility and elevating the threat of death when they lose their host; they can only survive for 1-3 days without feeding.

What science is involved in studying lice?

Minahil Syed

Summary: Approximately 1.9 million years, hominids began to migrate out of Africa, and archaic populations began to settle in Europe and Central Asia; these areas being were comparatively colder brought notice to their vulnerability towards the cooler temperatures (Toups, Kitchen, Light, Reed, 2011, p. 29). The transition from pre-homo sapiens to the anatomically modern human (AMH) facilitated numerous changes, but in the context of studying lice, the most prolific was the usage of clothing, which too is held responsible for the successful expansion out of Africa and ability to settle in colder regions with higher latitudes (Toups et al., 2011, p. 29). Determining when clothing use began is difficult as the material use, such being animal pelts and skins, would tend to degrade extremely quickly, erasing any proposed evidence which could possibly have been salvaged from the Late Pleistocene (began 2.6 million years ago and lasted until 117,000 years ago). As an alternative, lice are studied to determine when clothing use initially began amongst hominins, as the parasitic insects have cospeciated with their host (Toups et al., 2011, p. 29).

The louse demonstrates a parasitic symbiotic relationship between species wherein the parasites, or lice, live on or inside another subject, obtaining nutrition at the behest or from its host.

The most prevalent species of lice are those that gather on the head; they are an obligate species, meaning, they cannot survive apart from a human host (DeGrandpre, n.d.). Body lice differ from head lice insofar that their eggs are laid on clothing rather than directly on the body itself. Apart from the body, there also exist pubic lice that attach their eggs to the hairs – similar to head lice- wherein they abide (el-Showk, 2015). Resultant of their differing 'environments,' each species of louse has adopted a niche relative to their location on the body; by researching their evolutionary history and mitochondrial DNA, the time period wherein they diverged in can be deduced (el-Showk, 2015). It is between 30,000 and 110,000 years ago when modern human predecessors began wearing clothing that their genome reflected a variance in their particular role (el-Showk, 2015).

Head and Clothing Lice and their Primate Hosts

Head and clothing lice are closely related, sharing a common ancestry with chimpanzee lice. As suggested by natural historian 'Charles Darwin' and evolutionary biologist 'Thomas Henry Huxley,' and their following predecessors within the field of evolutionary science, humans and African great apes share common ancestors (Waterson, Lander, Wilson, 2005, p. 69). Contemporaneous molecular studies (e.g. protein-coding exons and genomic sequences of both humans and apes) have showcased that both the common chimpanzee (Pan troglodytes) and bonobo (Pan paniscus) are the closest humanistic variant amongst great apes (Waterson et al., 2005, p. 69); sharing a staggering 98-99.6% of DNA, often varying based on the relative study (Gibbons, 2012).

Protein-coding exons are parts of a gene that will encode a component of the mature RNA (ribonucleic acid) produced by that gene after introns (non-coding components of the RNA transcript) are removed via RNA splicing [a process wherein newly created precursor messenger RNA (pre-mRNA) is transformed into mature mRNA (Waterson et al., 2005, p. 69). It is during the splicing process that introns are removed and replaced by exons; which are then 'spliced' (fused) together instead ("RNA Splicing," 2014). The replacement of non-coding gene sequences (introns) by exons (protein-coding) hence occurs to allow the translation of mRNA into a protein, consequently increasing the diversity of mRNA expressed within the genome (Kelemen, Convertini, Zhang, Wen, Shen, Falaleeva, Stamm, 2012, pp. 1-3). Within eukaryotic cells (uni- or multicellular cells which are found in plants, animals, and fungi, and protozoa), introns are removed by a protein called a spliceosome, which is a ribonucleoprotein particle whose function primarily involves separation of transcripts and their association with a particular protein (Dreyfuss, Philipson, Mattaj, 1988, p. 1419). Had introns not been removed, the RNA would be translated into a non-functional protein instead (Harris, n.d.)

Whole genome sequencing (WGS) is a process which determines the sequence and order of an organism's DNA nucleotides (colloquially referred to as 'bases, including, Adenine, Guanine, Thymine, and Cytosine) ("DNA Sequencing Fact Sheet," n.d.). In order to sequence an entire genome, there is no presently available methodology; rather, scientists must break the genome into smaller components, sequence those pieces, and reassemble their relative order in order to revisit the entirety of the genome ("GENOME SEQUENC-ING," n.d.). The two primary approaches to DNA sequencing include the "clone-by-clone" approach and the "whole-genome shotgun" procedure.

Clone-by-clone sequencing involves 'breaking' the genome into 'bacterial artificial chromosomes'-sized pieces (BAC); a DNA construct which can be composed of thousands of base pairs used to incorporate genetic material into a target cell (Shizuya & Kouros-Mehr, 2001, pp. 26-29). BAC are used for transformation, which is the molecular alteration to a cell via the incorporation of surrounding exogenous genetic material and cloning, which is quite literally, the production of identical DNA via biotechnology (Shizuya & Kouros-Mehr, 2001, pp. 26-29). These BAC-sized chunks are called clones (approximately 150,000 base pairs long), using genome mapping techniques, the area wherein each clone resides within the genome is found, the clone is then sized into smaller parts for sequencing (approximately 500 base pairs), and since the order is based on the original strand, the complete sequence would form one complete chromosome at the end of the process (Nickle & Barrette-Ng, n.d.). Whole genome shotgun sequencing, although not as tedious as clone-by-clone sequencing, in that, it does not require the creation of a physical map before the procedure begins, faces the risk of redundancy amongst fragments when sequencing; the process breaks the genome into small pieces and then reassembles them by searching for overlaps in the sequence of each piece (Nickle & Barrette-Ng, n.d.).

DNA Sequencing

The process of DNA sequencing is used in order to calculate when clothing lice first began to genetically diverge from human head lice. In order to effectively determine human evolution and migratory patterns, scientists rely on DNA sequencing in order to calculate when clothing lice began to genetically and ecologically - in terms of an animal's distinctive niche - diverge from human head lice (Torrent, 2011).

Albeit the technology was initially developed in order to pinpoint when humans began to wear clothing, - which consisted of simple attire composed of loose animal skins, pelts, and hides- researchers noted that neither body nor clothing lice were a defined and separate category until clothing for regular use became a norm amongst human societies (Torrent, 2011). While geneticist Marx Stoneking, of the Max Planck Institute, estimated that humans began wearing clothing circa 107,000 years ago, results from the University of Florida study showcased newer research mechanisms and calculations in order to determine the closer estimate, tracing back further, 170,000 years earlier ("Lice DNA study shows humans first wore clothes 170,000 years ago," 2011). The motive being is theorized to have been to tackle the migration out of Africa into colder climates and higher latitudes,

49

Ice Age conditions prompted a need to stay warm in order to survive the journey away from home (Torrent, 2011). Although the last Ice Age occurred circa 120,000 years ago, data suggests that humans only began adorning their clothing in the Ice age preceding that, which occurred 60,000 years earlier (180,000 years ago in sum), based on temperature estimates resulting from ice core studies (Torrent, 2011). As a species, modern humans, homo sapiens, diverged 200,000 years ago (Torrent, 2011).

Unlike other parasites, lice are studied as they tend to acclimate to their hosts rather quickly, thus remaining stranded on their host lineages for long periods of evolutionary phases. The coevolution between host-parasite not only allows scientists to discover the evolutionary changes amongst the lice themselves (based on data noted on the parasite's changing niche, environment, and host), but the lice are likewise used as a marker for their hosts' evolutionary history as well. In reference to clothing lice, as stated by David Reed, curator at the Florida Museum of Natural History, "When new real estate opened up, new habitat became available…it only makes sense that head louse populations would try and make use of that habitat," Reed said. "And they were successful at it, that's why we have clothing lice today." (Torrent, 2011).

The Evolutionary History of Primate Lice

The coevolution of lice alongside their primate hosts has gone on for upwards of 25 million years; lice found on chimpanzees and that found on the human body/head share a common ancestor tracing back to nearly 6 million years ago, a separation reflective of the simultaneous differentiation amongst their relative hosts (Reed, Light, Allen, Kirchman, 2007, pp. 1-3).

Cophylogenetic analysis of the Pediculus (of the family Pediculidae, a genus of 'sucking lice') and Pthirus (a genus of lice of which two extant species have survived, of the family Pthiridae; Pthirus gorillae is found on gorillas, Pthirus pubis affects the pubic region of humans) (Reed et al., 2007, pp. 1-3) reveals that the duo are monophyletic, meaning that they are classified within the same taxon and share a common recent ancestor ("Concepts of monopoly, polyphyly, & paraphyly," 2012).

Phylogenetic and reconciliation analysis showcases that the present distribution of primate lice has two pathways ascribed to its current state, the one being that a Pthirus species switched from gorilla hosts to humans (which is why there is both a Pthirus gorillae and pubis), the other being that the divergence between Pediculus and Pthirus is resultant of the 'split' between gorillas and older lineage of chimpanzees (which, to recall, are the closest ape, alongside bonobos, in terms of genome, to contemporaneous humans) (Reed et al., 2007, pp. 1-4). Divergence dates between the Pediculus and Pthirus reveal that the coevolutionary histories of primates and their parasitic visitors has been graced by cospeciation, parasite duplication, extinction, and host switching amongst lice (Reed et al., 2007, p. 1). Using a mapping system called TreeMap, results aligned with the 'recent host switch' theory (that predicts that the differentiation between Pthirus pubis and Pthirus gorilla is more recent than the chimpanzee and human split) and 'pair of lost lice' hypothesis (which predicts duplication amongst louse which led to the ancestor lineages leading to the now-present genera of Pediculus and Pthirus) (Reed et al., 2007, p. 2).

Genome sequencing, tree mapping, and generalized analysis of the louse and their host switching explains not only how and why gorilla lice adapted to a completely different environment, such being human pubic hair, but also when human bodies began to change (resembling present-day hominids or homo sapiens), how coevolution between parasites and hosts occurs, and how parasites, specifically lice, diverged faster (14 times) than their hosts – a discovery attributed to mutations within the DNA (el-Showk, 2015).

The head louse (Pediculus humanis capitis) and body louse (Pediculus humanus corporis): Sequencing and Environment

The ecological differentiation between ectoparasites, P.h. capitis and P.h. corporis, likely occurred as a result of the adoption of clothing; while there is no direct archaeological evidence for such, DNA sequencing is used in order to ascertain and rightfully predict when this phase within human evolution occurred (Kittler, Kayser, Stoneking, 2003, pp. 1414-1415). Sequences were thus obtained from two mtDNA [(mitochondrial DNA); as opposed to nuclear DNA which is inherited from both parents, mitochondrial DNA is obtained only from the mother (allowing to trace material lineage)], two nuclear DNA fragments from a global sample of 40 body and head lice, and a singular louse from a chimpanzee to form the outgroup (Kittler et al., 2003, p. 1414).

The results establish that there is considerable diversity in African lice as opposed to their non-African counterparts while also suggesting that the origin of human lice can be traced to an African source (Kittler et al., 2003, p. 1414). Although humans began wearing clothing 170,000 years ago (based on the aforementioned UF figures), the molecular clock analysis within this study reveals that body lice appeared only 72,000 ± 42,000 years ago, while likewise suggesting that clothing is a relatively newer adoption within human development and growth.

Portions of the mtDNA ND4 (with 579 base pairs) and Cytochrome b (440 bp) genes were sequenced from 14 body 26 head and lice, having arrived from 12 spatially vast geographic regions (Kittler et al., 2003, p. 1414). MT-ND4 is found in human mitochondrial DNA and is one of the seven mitochondrial genes responsible for encoding the enzyme 'NADH dehydrogenase': a group of two or more polypeptide chains liable for respiratory

51

complex I (which is essentially responsible for maintaining function concerning aerobic energy metabolism) (Parey, Wirth, Vonck, Zickermann, 2020, pp. 1-9). CYTB, otherwise known as Cytochrome B, is a protein found in mitochondrial cells which functions as a component of the electron transport chain (wherein electrons are transferred from one molecule, donors, to another, acceptors, in order to eventually synthesize ATP, the principal molecule responsible for storing and moving energy from cells to fuel other cell processes) ("Electron Transport Chain," n.d.). Phylogenetic trees constructed for both the aforementioned ND4 and CYTB sequences were almost identical in terms of topology (the branching composition of the tree, the structure thereof shows interrelatedness amongst taxa) and length of branches (Kittler et al., 2003, p. 1415); while vertical lines represent when an evolutionary split occurred, branch lengths are a mechanism to posit how much genetic charge, divergence, and evolutionary time between two nodes has occurred or been noted ("Branches," n.d.). The topology of the three showcased that the deepest clades (a section within the tree that is composed of a common ancestor whom their lineal descendants emerge from) were made up of head lice sequences, which in turn means that body lice emerged from head lice, confirming the previous DNA sequencing/cloning theories in the previous sections (Kittler et al., 2003, p. 1414). It is integral to note, as all body lice sequences were absorbed under the guise or title of the combined head and body lice clade, this fact that there is no distinct separation or branching of the two species suggests ancestral polymorphism (Kittler et al., 2003, p. 1414). In essence, this means that in the past, there was no separation between head and body lice nor was there differentiation between their niches; hence, they took on more than their natural relative environment.

Unlike the comparatively aggressive estimates in the previous reports, a separate study comparing the origin of clothing lice with reference to anatomically modern humans in Africa uses a Bayesian coalescent modeling approach to date the emergence of clothing lice to a comparatively vast period between 83,000 to as early as 170,000 years ago. In genetics, coalescent theory assumes no recombination, natural selection, or gene flow amongst the populations, insinuating that each gene variant is as likely as another to have been passed down from one individual to another, hence the Bayesian method is proposed as a delimitation model for species using sequencing data (Zhang, Zhang, Zhu, Yang, 2011, p. 747).

Studying Lice Themselves

While the previous data generally suggests that body lice appeared as an offshoot from head lice when modern humans began wearing clothes, recent findings suggest that under conditions of extremely poor self-hygiene, a head lice infestation can turn into a massive infestation given the correct circumstances. In such circumstances, the head lice are able to ingest copious amounts of blood (a trait typical of body lice rather than head lice) allowing them to eventually colonize clothing as well (Amanzougaghene, Fenollar, Raoult, Mediannikov, 2020, pp. 2-3). Furthermore, several researchers note that under appropriate conditions (assuming environmental nutrition and temperatures are liable), head lice are able to develop into body lice ecotypes; an ecotype is a variant wherein phenotypic differences amongst species are too implicit to necessitate an entirely new sub-classification (Amanzougaghene et al., 2020, p. 3).

Conclusion

In researching the historical conditions, genetic processes, and mechanisms in which lice are studied showcases wherein variations of lice emerged from. Therein offering a perspective not only on ecotypes in modern species classifications but also concerning the evolution of pre-hominin/anatomically modern humans, migration out of Africa, the use and crafting of clothing in foundational societies, and the development of ecological niches amongst species of louse.

How does the environment affect lice?

Leah Heinen

Introduction

Lice are parasites that most often feed on blood or skin debris. They enjoy environments where hair is present so that they can be close to an organism's scalp. This chapter will discuss the preferred environments of lice on humans, as well as common pets like cats, and dogs. The chapter will also discuss how these environments could make an individual or an animal at a higher or lower risk of lice infestation. Hopefully, this chapter will help an individual to understand how to prevent or eradicate lice, as well as provide a more in-depth understanding regarding the relationship between lice and hygiene in the hope to break down the negative stigma surrounding lice infestation in society today.

Human-Specific Lice
Locations of Different Species

The most common species of lice found on humans is Pediculus humanus capitis or head lice (Rossini, Castillo, & González, 2007). However, humans can also carry Pediculus humanus corporis (body lice) and Pthirus pubis (crab/public lice) (Sanders & Stanhope). Head lice generally live on the hair of the head, and often reside close to the scalp, usually within 6mm (DSHS Texas). They are often found at the back of the neck and also behind the ears (DSHS Texas). Maintaining a close proximity to the scalp as well as residing in warm places like the back of the neck and behind the ears, allows lice to maintain their ideal body temperature (DSHS Texas). Body lice, on the other hand, prefer to remain on clothing or an individual's bodily hairs (Sanders & Stanhope). They will most often lay their eggs, also known as nits, on clothing (DeGrandpre & Marcin, 2019). However, body lice will travel to the body to feed (DeGrandpre & Marcin, 2019). Unfortunately, this type of lice may act as a vector for various diseases including louse-borne typhus, relapsing fever, and trench fever (please see chapter 3 for more details)(DeGrandpre & Marcin, 2019). Thankfully, current research and understanding concludes that this species of human lice is the only type of human lice that can spread disease (DeGrandpre & Marcin, 2019). Finally, crab lice are primarily found on hair that is found in the pubic areas of the human body(Sanders & Stanhope). However, crab lice can also be discovered on leg hair, armpit hair, or facial hair (Sanders & Stanhope). While these types of lice often do not act as vectors for disease, they provide much discomfort for their human host. This section of the chapter will mainly focus on head lice due to their prevalence and popularity, however, body and crab lice will also be briefly discussed when relevant.

Who Gets Head Lice?

Although it is important to understand that lice infestations can affect any human, some individuals may be at higher risk of coming into contact with lice, and therefore, are more at risk for infestation. For example, often individuals who live in crowded spaces are at higher risk of coming into contact with head lice (DeGrandpre & Marcin, 2019). For example, most commonly preschool, childcare, or elementary-aged children are those who have head lice infestation (DSHS Texas). Additionally, household members with children in these age groups are also more at risk due to exposure and close contact(DSHS Texas). These individuals may be more at risk for head lice infestation because, in environments such as preschools, daycares, and elementary schools, there is frequent close contact between children, making the spread of head lice possible and sometimes even unpreventable (DSHS Texas).

Human Lice and Hygiene

As opposed to common perception in society today, head lice do not prefer 'dirty' hair (CDC, 2019). In fact, head lice have no preference related to the cleanliness of an individual's hair (CDC, 2019). As discussed in previous chapters, head lice will feed off tiny amounts of blood from human scalps (DSHS Texas). Therefore, lice are not picky about the cleanliness of their new home, rather, they are primarily concerned with accessing blood as quickly as possible for survival and energy to reproduce(DSHS Texas). Interestingly, head lice that are not attached to humans (their host) will die in one to two days from starvation (DSHS Texas). This is why often when individuals are diagnosed with a lice infestation they are told to isolate items that cannot be washed such as brushes, pillows, and other items that may have lice on them for one to two days (CDC, 2019).

Although head lice do not prefer 'dirty' hair, conversely, body lice are more commonly found on individuals living in unclean environments (DeGrandpre & Marcin, 2019). Often, unwashed clothing is blamed for the spread of body lice (DeGrandpre & Marcin, 2019). This observation has led some to believe that body lice may be attracted to unhygienic environments. However, others may argue that the prevalence of body lice in uncleanly environments is primarily due to lack of washing clothes (and consequently getting rid of body lice) after contraction (DeGrandpre & Marcin, 2019). Whatever the cause, body lice can be easily prevented or controlled by regularly washing clothing items and by limiting sharing clothing items (DeGrandpre & Marcin, 2019).

Head Lice Hair Type Preferences

Ultimately, head lice prefer a habitat that allows them to abundantly reproduce and that allows them to feed (provides them with easy access to blood). Interestingly, certain hair types may provide more inviting environments for lice that meet their needs to a greater extent. The subsequent paragraphs will summarize which hair types are the most suitable environments for lice, and therefore, are the most susceptible to lice infestation.

A study was conducted in Norway regarding the incidence, predictors, and infestations of head lice (Birkemore et al., 2016). In this study, primary-school-aged children from 12 different schools in Oslo Norway were asked to complete a questionnaire regarding their hair type, sex, homelife, as well as various other "exposures" (Birkemore et al., 2016). This questionnaire was to be filled out by their caregivers and returned to their institution (Birkemore et al., 2016). Each child who completed the questionnaire was also screened for head lice (Birkemore et al., 2016). This study concluded the validity of various head lice preferences regarding hair characteristics such as colour, curliness, length, and thickness (Birkemore et al., 2016). For example, the study concluded that hair colour had no significant impact on lice preferences (Birkemore et al., 2016). Additionally, it was concluded that children with wavy hair had the highest incidence of lice, whereas individuals who possessed straight hair were least at risk (Birkemore et al., 2016). Furthermore, hair length was established in this study as one of the most significant characteristics of head lice infestation (Birkemore et al., 2016). This study concluded that individuals with medium-length hair had the highest incidence of lice compared to those who had either short or long hair (Birkemore et al., 2016). However, long-haired individuals were more likely to have lice than children with short hair (Birkemore et al., 2016). Finally, children with thick hair were more likely to have lice than those with thin hair (Birkemore et al., 2016). In summary, this study concluded that children with wavy, medium-length, thick hair are most at risk for lice infestation. Interestingly, this study also suggested that females were more at risk than males (Birkemore et al., 2016). However, these results could be attested to a female tendency to have long hair as this characteristic was identified as being one of the most predictive characteristics of head lice. The study also suggested that this risk could be attested to the fact that young females often have more frequent close contact with their peers than young males (Birkemore et al., 2016).

Lice infestation can affect all races and ethnicities (Guenther, 2021). However, in the United States of America, head lice infestation risk has been found to possibly be race dependent (CDC, 2019). For example, a head lice infestation is the least common in African Americans (CDC, 2019). It has been postulated that this may be due to the shape of head lice claws (CDC, 2019). These claws are shaped to more easily grasp certain shapes and various widths of hair (CDC, 2019). For example, American head lice claws can more easily grasp the round shape of the hair shaft of caucasian or Asian individuals (Guenther, 2021). Although African Americans may be at lower risk for head lice, sources state that they may still experience scalp infestation of crab lice (Guenther, 2021).

Other Preferred Environmental Characteristics

Head lice are versatile creatures that are not killed easily and can live in a variety of environments. This is one reason why head lice infestations are difficult to eradicate.

Temperature

Generally, head lice are not immensely affected by temperature unless these temperatures are extreme. For example, head lice and head nits will be killed at temperatures greater than 51 degrees Celsius or 125 degrees Fahrenheit (CDC, 2020). Head lice and head nits can also be killed at extremely low temperatures. For example, the species will die if exposed for over 48 hours to a temperature of 17 degrees Celsius or 0 degrees Fahrenheit (CDC, 2020). This is why when lice infestation is diagnosed, it is suggested that an individual wash and dry items they were in contact with such as their clothing and bedding in the hopes that the heat from the dryer will eradicate all living lice on these items (CDC, 2020).

Wet and Humid Environments

Generally, lice thrive in environments similar to that of one close to the human scalp which in most cases is warm and humid (CDC, 2020). However, studies have concluded that head lice can survive underwater for many hours (CDC, 2020). Researchers who have studied this phenomenon observed head lice holding tightly to human hair when submerged underwater (CDC, 2020). This may be why sources state that it is unlikely to spread head lice by swimming in a pool with another individual who has a head lice infestation (CDC, 2020). Although head lice can survive in water for a great amount of time, as previously mentioned, they thrive in warm humid conditions (CDC, 2020). More specifically, nits must incubate in the warm humid environment that a human scalp would provide (CDC, 2020). This is why most nits are found very close to the scalp (CDC, 2020).

Oils and Scents

Various oils and scents deter lice, and therefore, by using or wearing these products, a non-ideal environment for lice is created, thereby reducing one's chance of lice infestation. For example, an environment with tea tree oil may not be preferred by lice. Many studies suggest that tea tree oil could be used preventatively to stop head lice infestations. However, more research is needed before strong conclusions can be made. A study from the International Journal of Dermatology looked at the effectiveness of tea tree oil, lavender oil, peppermint, and DEET for preventing lice infestation in primary school children (Canyon & Speare, 2007). This study concluded that tea tree oil was the most effective oil for preventing lice (Canyon & Speare, 2007). Interestingly enough, tea tree oil can also be used to treat lice infestation. This fact was proven by a randomized controlled trial (RCT). This RCT found that 97.6% of children with lice that were treated with a tea tree oil and lavender mixture were louse-free at the end of the study (Barker & Altman, 2010). However, only 25% of children with lice that were treated with commonly used products (Pyrethrins-based) were louse-free at the end of the study (Barker & Altman, 2010). Therefore, using a tea-tree oil and lavender mixture, may be a solution to the problem of "super lice" that will be discussed in greater detail in Chapter 12 of this book. Additionally, mango-scent has been observed to prevent head lice infestation. Dr. Hannah Chow-Johnson, who is a pediatrician at Loyola University Health System, stated that in addition to tea tree oil, mango and rosemary are scents that help prevent infestation as well (Loyola University Health System, 2012). However, extensive studies have not been conducted on the preventative effects of mango scent and so this suggestion should be taken with some degree of caution. Rosemary, however, has been studied as a possibly effective approach to prevent head and body louse(Rossini, Castillo, & González, 2007). Furthermore, other essential oils that create unpreferred environments for head lice include juniper, lavender, geranium, lemon, rose, cinnamon, lemongrass, thyme, myrtle, oregano, and eucalyptus (Rossini, Castillo, & González, 2007).

Lice that affects Animals

Often, when fall approaches and the weather starts to cool, lice infestations will start to become more prevalent (Washington State University). Understanding the preferred environments of lice in animals can allow animal owners to ensure the safety and health of their pets (Washington State University). Interestingly, lice are species-specific, meaning that in most cases each animal species can be infected by only their own set of lice species (Washington State University). Many animals have the ability to be infested with lice including sheep, horses, cattle, swine, cats, dogs, etc.(Washington State University). However, this chapter will focus on only cats and dogs and the preferred environments of lice that infect these specific animals because they are the most common animals that humans come in contact with.

Cat and dog lice are very different from human lice. For example, they tend not to move around as much as human lice and instead, they stay in one place at the base of the animal's fur (Wondra, 2021). Additionally, although human lice do not pose a serious threat to the spread of disease, cat and dog lice can carry and transmit various diseases (ie. Tapeworms) (Irwin & Jefferies, 2004).

Dog-specific Lice
Different Species

Dogs can be infested with three different types of lice (Thomas, 2018).These include Linognathus setosus which is known as the bloodsucking louse, Trichodectes canis which is also referred to as a biting louse, and heterodoxus spiniger which is a biting louse that feeds on a dogs blood (Thomas, 2018). The Heterodoxus spiniger species, however, is a very rarely found type of lice in North America (Thomas, 2018).

Where do they get lice?

Unfortunately, as opposed to human lice, lice that attack dogs generally infect animals in poor health or who live in unsanitary areas leading to the understanding that dog lice may be attracted to 'dirty' environmental conditions (Thomas, 2018). However, the transmission of lice in dogs is usually brought about through the direct contact of a dog with another lice-infested dog (Thomas, 2018). This means that doggy daycare centers, shows, kennels, and parks may be environments where dogs could be infested (Thomas, 2018).

Seasons and Temperature

Interestingly enough, biting louse infestations are more commonly found in the winter and the spring, than in the summer (Benelli, Caselli, Di Giuseppe, & Canale, 2018). Studies suggest that the change and the fluctuation of risk could be due to high summer temperatures and solar radiation (Benelli, Caselli, Di Giuseppe, & Canale, 2018). Just like human head lice, dog lice are also damaged or killed by high temperatures (Thomas, 2018). This is why an important prevention strategy to prevent lice infestation in dogs is known as 'environmental cleaning' (Thomas, 2018). Environmental cleaning involves conducting high-temperature cleans of bedding, rugs, brushes, and other items that the dog commonly uses to kill any living lice and prevent infestation (Thomas, 2018).

Cat- specific Lice
Different Species

There is only one cat-specific species of lice which is Felicola Subrostratus, a feline chewing louse (Thomas, 2018). Interestingly, this species can be seen with the naked eye (Thomas, 2018). As opposed to only a one to two day survival without a host in humans, this species of lice can live up to ten days without its host (Grant, 1989). Unfortunately, cats who are infested with lice may contract various skin diseases from this species of lice as well as spread disease (Thomas, 2018).

Where do they get lice?

Just like dogs, cats generally are infested with lice because they are living in poor, dirty conditions whether indoors or outdoors leading to the suspicion that this form of lice prefers 'dirty' environments (Thomas, 2018). Additionally, cats can spread lice through direct or indirect contact (Thomas, 2018). Direct contact involves cats rubbing up against each other or being in close proximity with one another. Direct contact may occur if there are multiple cats in one home, cat-sitters, or even if outdoor cats rub up against one another. Indirect contact is the spread of lice through inanimate objects such as a brush or a bed that two cats may share. Because feline chewing lice can live up to ten days without their host, the indirect spread of lice is much more common with cats than with humans.

Effects of Temperature

Just like human and dog lice, feline chewing lice are very temperature sensitive and prefer warm, humid environments (Thomas, 2018). This is why in temperate regions lice are more common in the colder months than in the summer (Thomas, 2018). Therefore, it is recommended for cat lice as well to do environmental cleaning when a cat contracts lice to prevent continued infestation (See "Dog Lice, Seasons and Temperature" for more information)(Thomas, 2018).

Chapter Summary

In conclusion, just like many organisms, lice focus on feeding and reproducing. To do these things, they must find a suitable environment. In humans, they look for thick, medium-length, wavy hair. They will be attracted to environments that are not too hot nor too cold. Additionally, they will be deterred by environments with certain oils and scents such as tea tree oil, lavender, rosemary, and mango. As discussed previously, lice cannot fly or "jump" meaning that they are often found in populations that have frequent close contact with other individuals, making children who attend daycares and schools more susceptible to lice infestation. Lice that reside on dogs and cats have similar environmental preferences. However, as opposed to humans, they often will infest animals who are "dirty" and live in poor conditions. These types of lice also prefer environments that are not "too hot" or "too cold" and because of this, they will often be more prevalent in the cooler months of the year. Dogs and cats who have close contact with other animals of the same species are more at risk of lice infestation. Because cat lice can live up to 10 days without a live host, cats who live with many other cats are more susceptible to infestation because lice can be more easily spread through inanimate objects such as shared brushes or beds.

How are lice portrayed in popular culture?

Ipsa Gusain

Every child has heard of the dreadful lice experience; they have either directly, or indirectly encountered the invasive insects. Lice are infamous for their infectious nature regardless of how relatively harmless they are. Throughout this chapter, the portrayal of these parasites in popular culture will be discussed, as well as the truth behind many popular myths and where these stories originate from.

Popular Beliefs: Head Lice

Countries all over the world have developed a fear of obtaining lice. This is not due to the perceived dangers of such insects, rather the social stigma and the underlying fear of rejection that exists amongst those that have been infected. Another crucial factor that feeds into this social ostracization is the misinformation spread by family members, media, and within school settings.

The following paragraphs discuss the many ill-informed beliefs about lice spread by society. Many people have early memories of a classmate contracting head lice and being sent home due to the infectious agent and others have experienced this first hand. This forced quarantine by many school boards across many countries, especially in the Western world, is done so that the lice cannot infect other students due to the close-contact nature of these institutions. Children, predominantly under the age of 12, use school as a way to interact with friends and fulfill their social needs. For most boys and girls, this includes

62

close contact with others in the form of playing games like tag, human knot, hide and seek etc. Other forms of contact may include sports and sharing objects such as hats and sports equipment like jerseys. Overall, most people see school as the main location for the transmission of head lice.

There seems to be this collective misconception that lice can jump from head-to-head and dwell in objects until they are transferred over to another human head to feast on their blood. For most girls, many are warned time and time again, not to share hair brushes, hair accessories, scarfs, combs, towels and hats with others to prevent getting head lice. The fear of these spider-like insects forces young kids to adopt the habit of changing and washing sheets and clothes regularly early on as another preventative measure. In fact, many go as far as to quarantine or throw out any item that has come in contact with the person perceived as having lice.

Preventative measures predicated on the notion of cleanliness festers a culture of shame; this idea that the individual must be shunned by their community immediately after being caught with head lice. This is such an effective method of instilling a fear of lice within society, that these individuals feel disgusted with themselves for not practising good hygiene. The humiliation and teasing of a person caught with having lice is so extreme, that Hurst et al. (2020) reports a decrease in self-perception within these children. Having lice, as generally believed by society, is due to socially constructed ideas regarding cleanliness, parental responsibilities, health, and stereotypes based on class. With lice so closely tied to ideas about cleanliness, most people feel justified in using the blame-and-shame approach. Many perceive those they suspect of having lice as being dirty or having bad hygiene and by association, due to stereotypes, being lower class or having a lower socioeconomic status. There is an interesting link between class standings and the history of lice; as the living standards of people increase, so does the social stigma around lice. Even now, people associate having lice with homeless or refugee populations. This sense of shame and embarrassment is also seen in the parents of children found with lice; in a survey done by Silva et al. (2008), 92 percent of parents reported feeling embarrassed when their kid was infested with lice. Parents were aware of how their child was being perceived in the school setting but over 90 percent were too afraid to seek professional help and instead, chose to self-medicate with a louse comb (Hurst et al., 2020; Silva et al., 2008). Unfortunately, head lice is still common but due to the shame-and-blame approach, most families keep the infestation hidden from the public. Therefore, most individuals believe this is a rare occurrence which in turn, makes it even more difficult to talk about.

Lastly, the social stigma around having lice urges parents to remind their kids what to avoid doing, in order to avoid the humiliation and stereotypes that usually accompany lice infestation. Avoidance techniques include not scratching one's head in public as itching is seen as the main symptom of lice, staying away from those perceived as having lice, and as discussed previously, not sharing items in close contact with the head. As a consequence, many kids are taught from a young age the stigma around lice and cleanliness, rather than the truth regarding transmission and contraction.

These popular beliefs, albeit not entirely rooted in truth, allow a culture of shame, secrecy and isolation to thrive when the best way to tackle mass infestation is through communication as well as frequent and honest lice check-ups. But what is the truth behind these common lice myths?

Myths Busted

Head lice have been the target of much hearsay over the centuries. The spread of countless rumours and beliefs about head lice, led to many myths on transmission, symp-toms, treatments and the true dangers of the blood-thirsty parasites. The aim of debunking some of these popular myths regarding lice is to dismantle the stigma around head lice and create open communication amongst communities as this is the most effective way of protecting children from lice infestations.

There is a common misconception within the general public that transmission of head lice is very simple and easy; most people imagine head lice to be tiny insects that have the ability to jump from head-to-head which makes any close-contact environment a breeding ground. Schools are popularly believed to be setting where most lice infestations thrive with children being the main targets. There is some truth to the theory that children are most likely to be infected with head lice; the Centers for Disease Control states that up to 12 million children are found with lice yearly (Hurst et al., 2020). However, head lice do not often spread in a school setting because it is, contrary to popular beliefs, not the ideal setting for proliferation (Hurst et al., 2020). This due to head lice needing direct head-to-head contact. They do not have the ability to survive off of inanimate objects as human blood is their only means of survival (Hurst et al., 2020). Lice must stay close to the human scalp as it provides food and shelter and as a result, they cling to hair strands tightly, never leaving on their own accord (Nash, 2003). Head lice are classified as parasites and therefore need to feed off the human host to survive, even at the expense of the host. Therefore, the widely spread warnings against sharing hats, hair accessories, towels, beddings etc. are deemed false and ineffective. This also means that school policies which force students to leave the premises until they are lice free, are also ineffective and quite frankly, harmful to the child. Policies discriminating against these students exists in over 80 percent of U.S. schools (Hurst et al., 2020). All these policies do is reinforce stigma, lower self-esteem and embarrass families. Exclusion-based policies are not recommended as the child does not benefit from isolation and misses out on education and community activities. These social beliefs are predicated on society's misinformed and outdated ideas on lice transmission, contagion, and prevention strategies.

The public image of someone with head lice is often a young girl itching her scalp and showing signs of extreme discomfort. This is a symptom of lice infestation, however it is not as common as many would believe. What is popularly believed to be one of the only discernible symptoms, other than seeing the lice themselves, is actually a widespread myth. Lice infestation is usually asymptomatic (Hemond, 2012). When a louse resides in one's scalp, it will bite at the sensitive skin to get to the blood. The bite is then surrounded by the lice's saliva which may cause itchiness due to an allergic reaction to the saliva (Hemond, 2012). While many children do scratch and itch from lice, 50 percent do not (Silva et al., 2008). While this makes itching the most common symptom, it is not the most promising way to determine pediculosis, or lice infestation. The best method for parents and guardians to keep their child healthy and free from lice, is to frequently check for lice.

While transmission and symptom confusion are often the topic of lice mythology, some well-known treatment alternatives are also fabricated and ineffective, if not borderline dangerous. Many substances over the years have been passed down as home remedies including vinegar, alcohol, bleach and various oily substances (Silva et al., 2008). Unfortunately, while having no effect on the lice infestation, some substances should not be used on open wounds especially not on the sensitive skin of the scalp. The most effective way to get rid of pediculosis is with a louse comb or a fine-tooth comb. These wingless insects are very persistent and take a lot of time and dedication to eliminate (Hemond, 2012). While medical assistance does exist for lice infestation, most parents decide to deal with the problem themselves to avoid the stigma that comes with asking for help.

One of the most common myths one will hear about pediculosis is that they only infect those with an unhygienic lifestyle, whether that be due to class, socioeconomic status or parenting. It is insinuated, through exclusionary policies, that those perceived to have lice allowed the insects to invade their person and should be ashamed of themselves. The myth that those with pediculosis are dirty, poor or from developing countries still exist to this day. According to Hurst et al. (2020), there is evidence that head lice do not only affect the poor or those that have poor hygiene. Head lice do not discriminate based on cleanliness, rather direct contact is the primary method of proliferation. In most well-developed countries, lice are seen as abhorrent and thus, as discussed previously, the victims of these parasites are also treated poorly within society (Hurst et al., 2020). Some ascribe the characteristics of the infamous body lice to the relatively harmless head lice. The ambiguity of the term lice, is used as a way to portray head lice as dangerous and simultaneously, to justify the treatment of the people infected.This will be further explained in the next section however, head lice are low-risk parasites that are really just tedious to eliminate; they are not dangerous to human health (Hurst et al., 2020). Although the chance is low, head lice could transfer other pathogens, affect sleep due to the excessive itching and bleeding and/or cause anemia due to iron deficiency (Silva et al., 2008). Once again, these conditions are rare and pediculosis are not recognized as a medical disease (Silva et al., 2008).

These common myths regarding head lice are important to combat because they have led to much pain within families and caused a lot of isolation in kids and adults of all ages. The negative stigma surrounding head lice is attributable to the other forms of lice: pubic and the severe and deadly, body lice.

Other Types of Lice and Typhus

Pubic lice are another lice species that has peaked the interest of the public. This unique insect is highly sedentary and seeks a home in pubic hair (Anderson & Chaney, 2009). They spread through sexual contact and are considered a sexually transmitted disease (STD). In a study done by Anderson & Chaney (2009), similar to the head louse, pubic louse is also popularly believed to jump around until it finds a host. The general public believes pubic lice to proliferate not only through direct sexual contact, but also through toilet seats and clothing (Anderson & Chaney, 2009). Fortunately, due to their sedentary lifestyle, transmission through objects are highly rare and lice do not often leave the human body. Since pubic lice are not closely associated with bad hygiene, one can note the absence of any blame-inducing stigma. However, having pubic lice is still an STD with its own history of poor social management. Up until the twenty-first century, any form of STD was negatively viewed by the public; it symbolized sexual impurity, irresponsibility and highlighted the double standards between a man and a woman engaging in sexual activity. Now, pubic lice are one of the least stigmatized STDs amongst the following seven: HIV, syphilis, gonorrhea, genital warts, genital herpes and chlamydia. The increased advertisement to raise awareness of the dangers of HIV has made STDs like pubic lice less fear and shame-inducing (Anderson & Chaney, 2009). The history and public perception of pubic lice closely resembles that of head lice, but the recent removal of the shame and stigma surrounding it allows individuals to better handle an onset of this lice.

The last form of lice, offers insight to why society holds its current beliefs about head lice. The body lice is the most dangerous type; it has a history of death and destruction and has been revered for centuries before slowly declining. The years after the second World War marked its decline in the Western world and it is now considered a disease of the past (Bechah et al., 2008). Hans Zinsser's 1935 classic, Rats, Lice and History, recounts the way body lice has influenced the world through epidemic typhus (Weissmann, 2005). Zimmer portrays body lice as creatures that "lurk in dark corners and stalk us in the bodies of rats, mice and all kinds of domestic animals; which fly and crawl with the insects and waylay us in our food and drink and even in our love" (Weissmann, 2005, p. 492). The book became an international success in its depiction of the dreadful infestation; it also showed how many people around the world related to the harrowing details of body lice (Weissmann, 2005). This plague devastated empires such as Justinian's Byzantine and stopped Napoleon's Grand Army from invading Russia by affecting one-third of his soldiers during the great retreat (Bechah et al., 2008; Weissmann, 2005). Typhus also obliterated populations all over the world; some of the regions most notably affected in the twentieth century alone, included Northern Africa, Southern Italy and central as well as Eastern Europe (Bechah et al., 2008). During the second Great War, concentration camps also saw deadly outbreaks, further contributing to the astonishing mortality rates. The latest case of epidemic typhus occurred in 1997, Burundi, amongst the refugee population displaced due to the civil war where 100 000 people were overtaken with body lice (Bechah et al., 2008; Fournier et al., 2002). Typhus thrives during wartime where poverty, famine, negligence, and over-crowding are present (Fournier et al., 2002; Weissmann, 2005). Zinsser's book serves as a reminder of the devastating legacy left behind by body lice in the long history of humanity.

Over the years, the threat of body lice has lessened and in some areas of the world, completely disappeared. However, the impact still exists today in the form of misinfor-mation; people have displaced the truths of the near-extinct body lice, with the myths and legends of today's head lice. The key difference between the two types is that body lice have the capacity to carry deadly diseases such as typhus, while head lice may cause severe itching. Families have passed down the fear of body lice from generation to genera-tion and somewhere along the way, people stopped differentiating between these parasitic sister insects (Hemond, 2012). Epidemic typhus is rampant in unhygienic settings such as crowded locations in poverty-ridden neighbourhoods and during wartime (Portillo et al., 2015). Symptoms include rashes, high fever and headaches (Bechah et al., 2008). Further, body lice depend on ideal, low temperatures and cold months (Bechah et al., 2008). In fact, the lice were shown to leave the host if they became too hot due to relapsing fever (Bechah et al., 2008). One preventative method was washing one's clothes at 50 degrees celsius as body lice and eggs reside in the clothing of their hosts (Bechah et al., 2008). As previously mentioned, typhus has been on the rapid decline since the twentieth century however, this disease is re-emerging within certain homeless groups in developing countries (De Liberato et al., 2019). Unlike head lice, body lice lay eggs on the seams of clothing where they hatch if clothes are not kept clean and the host shows poor hygienic practices (De Liberato et al., 2019). Body lice have also been known to carry other pathogens such as Borrelia recurrentis which causes relapsing fever, and Bartonella quintana which causes trench fever (De Liberato et al., 2019). The spread of disease does not occur directly by the insect bites, but rather from the contamination of the site (Bechah et al., 2008). Overall, body lice are associated with high incidence of disease with the disease being fatal in up to 40 percent of the human hosts (Fournier et al., 2002). Thus, the fear of lice was instilled in every person alive during the periods where epidemic typhus devastated communities.

The parallel between body and head lice show a trend. Body lice spread by infected clothing, carry fatal diseases, thrive in unsanitary settings and predominantly affect the poor, refugees, and other people from lower socioeconomic status. Therefore, while the general public believes many falsehoods regarding head lice, all the myths are rooted in truth. Body lice has had such a catastrophic effect on human history, that head lice stand today as a reminder of human frailty while reinforcing cautionary behaviour, even when it is unnecessary such as the case with head lice.

What research is being done into lice and what future discoveries are possible?

Grace Parish

Introduction

Head lice, or Pediculus humanus capitis, is the most common form of lice. Head lice affects school-aged children across the world (Galassi et al., 2021). Lice are extremely common among this group, spreading by touching heads and close physical contacts. A 2001 study at the Hebrew University of Jerusalem found that upward of 5 percent of all children tested were infested with pediculosis (Weintraub, 2017). Some studies have cited this as high as 20 percent (Weintraub, 2017). Lice can also spread as body lice, or Pediculus humanus humanus (Weintraub, 2017). Body lice can cause further concerns, proliferating diseases like epidemic typhus and trench fever (Weintraub, 2017). There exist additional under-discussed species of lice, such as sea lice, holding serious implications for disease transmission, agriculture, and fishing. Most urgently, the growing issue of lice resistance to insecticide treatment causes some concern as to our future ability to control the spread of lice and associated microorganisms. Luckily, this danger may also present great opportunities to adapt treatments to become more resilient, easy to produce, and more natural for use on young patients. Further attention and development of new methods to both model and treat different ecotypes of lice will reap significant benefits for society and economic success.

Lice and the scientific community

Previous scientific research on lice has pertained overwhelmingly to the two most common ecotypes: head lice and body lice (Amanzougaghene et al., 2020). Differing in morphology and biology, studies have shown that both body lice and head lice alike can transmit disease and pose serious implications for public health (Amanzougaghene et al., 2020). Ongoing research in the field focuses on limiting the spread of lice, preventing the development of treatment resistance in head lice, and eliminating lice altogether (Amanzougaghene et al., 2020). Head lice and body lice are almost genetically identical and affected similarly by origin, geography, and environmental signals (Amanzougaghene et al., 2020). Head lice live, breed, and lay eggs on hair shafts and feed on human blood every four to eight hours (Amanzougaghene et al., 2020). Conversely, body lice live, breed, and lay eggs in clothing, feeding less frequently but taking in greater quantities of blood than head lice (Amanzougaghene et al., 2020). Body lice are also more resilient in harsh environmental conditions. They can survive in lower humidity and for longer than 72 hours off of a host (Amanzougaghene et al., 2020). The variety in lice ecotypes and their differing behaviour based on environment provides a wealth of areas for future research and investigation within the scientific, public health, disease transmission, and patient treatment fields (Amanzougaghene et al., 2020).

One particularly important species to parasite research is human lice, or hemimetabolous (suborder: Anoplura, order: Phthiraptera), commonly referred to as sucking lice (Amanzougaghene et al., 2020). Sucking lice feed on human blood, using their piercing mouths to bite their hosts (Amanzougaghene et al., 2020). This mechanism lends itself to disease transmission in many ecotypes of lice. Lice can be infected by taking in the blood of an infected host, then carry this infection to a new, uninfected host through bite sites and microlesions on their skin (Amanzougaghene et al., 2020). For example, the species Rickettsia prowazekii facilitated the spread of epidemic typhus, necessitating antibiotic treatments and posing a recurring health threat to patients and society (Amanzougaghene et al., 2020). Most prominently, body lice have been observed to carry and transmit pathogens bacteria, as well as absorb microorganisms like viruses and hemoparasites (Amanzougaghene et al., 2020). The spread of microorganisms via body lice is of particular interest for future research (Amanzougaghene et al., 2020).

There are additional implications for analyzing the sustainability and the aquaculture industry to be found from studying sea lice (Government of Canada, 2019). Sea lice are a species that live in water, feeding on mucus, tissue, and blood of host fish (Government of Canada, 2019). The abundance of sea lice has begun to threaten salmon populations with the impacts of climate change on ecosystems; the increasing closeness between sea lice and the typically separated adult salmon and juvenile wild salmon has spread sea lice to the juvenile variety, proving lethal to their thin skin and small size (Carroll, 2008). Governments will need to implement measures to address the impacts of sea lice to maintain tourism and fishing industries (Carroll, 2008). According to ongoing research and innovation efforts, this may include the use of physical light traps and biological filters to limit and track sea lice spread in commercial settings and remove adult and planktonic sea lice from the ecosystem (Government of Canada, 2019). The use of cunners as "cleaner fish" to remove sea lice in Atlantic salmon farm cages is also being applied in Newfoundland, Canada (Government of Canada, 2019). A similar process may be possible using filter-feeding shellfish species like blue mussel and Japanese Scallops to consume sea lice (to date only tested in lab conditions) (Government of Canada, 2019). However, some sea lice treatments may harm non-target organisms, as is being studied in New Brunswick, Canada. Chemicals like AlphaMax and Salmosan treat sea lice but with long-term exposure may harm larval, lobster, and shrimp in the ecosystem (Government of Canada, 2019). Interestingly, an overarching and growing area of research supporting each of these trials is the development of technology and computers to model sea lice dispersal and behaviour, along with velocity, salinity, and temperature in their environment (Government of Canada, 2019).

Treatment market

Blood sucking lice have been around for at least 25 million years (Willingham, 2011). However, the widespread use of pesticide treatment for lice infestation dates back only to World War II, when the insecticide DDT was applied as a treatment on millions of people in Europe and Asia to prevent body lice (Weintraub, 2017). The use of insecticide DDT was used for decades to come until the 1980s when it was phased out due to safety concerns (Weintraub, 2017). Later, with the introduction of pyrethroids to treat lice, treatments were initially observed to be highly effective (Weintraub, 2017). Just one treatment would destroy all adult lice and even remaining eggs on one's scalp within weeks (Weintraub, 2017). Then, within just a few years, treatments for lice with pyrethroids were observed to be much less effective and unfortunately were not killing the majority of lice present (Weintraub, 2017).

Though chemical insecticides remain the most popular treatment method, other therapeutic options include topically applied physical agents, herbal recipes, and mechanical methods like combs (Amanzougaghene et al., 2020). There is growing demand for natural products and treatments. Recently, pediculicides made from plant-derived essential oils like eucalyptus and tea tree oils have gained traction (Amanzougaghene et al., 2020). However, further testing remains to be done to clinically evaluate their effectiveness (Amanzougaghene et al., 2020).

Treatment-resistant lice

The long trusted convention head lice treatments, over-the-counter shampoos with pesticides in the form of pyrethrins or pyrethroids (synthetic pyrethrins), have become for the most part ineffective (Weintraub, 2017). The ineffectiveness of these treatments is due to the widespread development of resistance to the chemicals by lice. This phenomenon is similar to that of antibiotic resistance, any lice able to survive treatment and continue to thrive where small amounts of the chemical remain present grows immune to the applied treatment (Weintraub, 2017). A toxicology study at the University of Massachusetts Amherst found that between two-thirds and three-quarters of head lice are indeed immune to the effects of pyrethrin and pyrethroid treatments (Weintraub, 2017). This is especially troubling to the majority of those infected, that is school-aged children, who are typically required to stay home until their scalps are free of both lice and nits (lice eggs) (Weintraub, 2017).

The widespread use of pesticide treatments like insecticide DDT have caused lice to develop resistance (Weintraub, 2017). DDT functions by disrupting a louse's nervous system, entering the cell membranes of nerve cells by increasing the flow of sodium ions into the cell (Weintraub, 2017). This causes the nerve cells to fire uncontrollably, paralyzing and killing the louse (Weintraub, 2017). Resistance to these treatments develops when there is an incomplete kill of the lice and some level of the chemical treatment remains on the scalp (Greive & Barnes, 2017). Over time and generations of exposure to these treatments, lice has evolved to block treatment and reduce the influx of sodium (Weintraub, 2017). The sample principle of altering the flow of sodium ions also applies to pyrethrins and pyrethroid treatments (Weintraub, 2017).

New treatments

Continued exposure to the same pesticide based treatments over the last several decades has built resistance in lice to all sodium flow disturbing approaches (Weintraub, 2017). Therefore, to fully defeat and eventually eliminate lice, novel treatments must be developed. For example, recently in Europe, some nonpesticide treatments are being researched and tested with success (Weintraub, 2017). In the United States, efforts are underway to incorporate new prescription medicines as treatments, with special consideration to preventing the development of further drug resistance (Weintraub, 2017). To date, the US Food and

Drug Administration has approved the use of three prescription drugs to treat lice: Ulesfia containing high levels of alcohol to suffocate lice, Natroba over activating nerve cells, and Sklice blocking nerve signals (Weintraub, 2017). To maintain the effectiveness of the new strategies and prevent similar resistance as previously seen, these three treatment options should be rotated and varied among patients to avoid overexposing lice to any single drug; however, research and guidelines for health care professionals on best practices for distribution have yet to be fleshed out and standardized (Weintraub, 2017).

The increase in head lice resistance to insecticide treatments, along with public concern over the safety of using these chemicals on young children, is driving interest in essential oil based treatments (Greive & Barnes, 2017). Many parents and pediatricians have long called for more naturally derived treatments to be considered and made more widely available in a commercial setting (Greive & Barnes, 2017). Plant oil treatments are the subject of further research as to their effectiveness. They may serve as a more effective, as well as more easily extractable and biodegradable alternative to the increasingly ineffective insecticide based chemical treatments (Greive & Barnes, 2017). In 2017, an Australian study assessed the safety and efficacy of a eucalyptus oil and Leptospermum petersonii based topical treatment in children with live head lice (Greive & Barnes, 2017). The treatment was observed to be more than twice as effective on participants after a single application as the standard insecticide treatment with limited adverse effects (Greive & Barnes, 2017). In vitro exposure of the same solution to lice and eggs was also successful, resulting in 100 percent mortality (Greive & Barnes, 2017). Experimentation using eucalyptus oil as well as other natural plant oils like lavender oil, wild bergamot, clove, tea tree, and Yunnan verbena have long been considered and practiced as household remedies (Candy et al., 2017). Moving forward, investing into clinical research and approval of said essential oil derived treatments may lead to a drastic improvement in both safety and effectiveness of commercial head lice treatments globally (Candy et al., 2017).

COVID-19

In March of 2020, the World Health Organization (WHO) classified the global spread of COVID-19 as a pandemic (Galassi et al., 2021). With an immediate transition to social distancing and self-isolation, in-person activities and congregating were halted. Non-essential workers largely took measures to work remotely, and schooling for children was interrupted or conducted virtually. This disruption of direct and close contact between children initiated an obstruction in the circulation pathways of headlice, changing the evolutionary dynamics of head louse populations (Galassi et al., 2021). During this time, the Argentina Centro de Investigaciones de Plagas e Insecticidas conducted a cross-sectional descriptive study via questionnaires to families in the metropolitan area of Buenos Aires, Argentina during the strictest 180-day lockdown period (Galassi et al., 20210). The Argentinean survey studied 1118 children and found that the prevalence of head lice infestation decreased by 25.7% during the COVID-19 lockdown compared to prior (Galassi et al., 2021). The distance and spatial distribution of hosts influences the presence of parasites, including head lice (Galassi et al., 2021). For the wider health research community, the COVID-19 lockdown allowed for unprecedented analysis of populations like head lice in a new and dynamic setting (Galassi et al., 2021). This ability has and will continue to provide valuable insight into preventative measures for the future (Galassi et al., 2021).

Summary, future discoveries

The presence and spread of lice is a widespread global issue. Over the past decade, the body of knowledge around lice and their genome has greatly expanded; however, there are many aspects to be investigated further. Pressing areas of concern are becoming more serious with lice resistance to insecticide chemical treatments. Additional research is also being directed towards achieving a greater understanding of the history of human evolution by the presence or absence of host fossils and DNA (Amanzougaghene et al., 2020). For example, body lice, which reside in human clothing, could only have evolved to their near current state after humans had worn clothes (Amanzougaghene et al., 2020). Furthermore, medical interest has been expressed in regards to research on the relationship between lice and other human host parasites as well as implications for agriculture and the environment (Amanzougaghene et al., 2020). The future implications of lice reach far beyond the infection of school-aged children with head lice, and have the potential to unlock unprecedented knowledge of genomics, disease control, and innovation.

References

Chapter 1

Amanzougaghene, N., Fenollar, F., Raoult, D., & Mediannikov, O. (2020). Where Are We With Human Lice? A Review of the Current State of Knowledge. Frontiers In Cellular And Infection Microbiology, 9. https://doi.org/10.3389/fcimb.2019.00474

Bonilla, D., Durden, L., Eremeeva, M., & Dasch, G. (2013). The Biology and Taxonomy of Head and Body Lice—Implications for Louse-Borne Disease Prevention. Plos Pathogens, 9(11), e1003724. https://doi.org/10.1371/journal.ppat.1003724

Boutellis, A., Abi-Rached, L., & Raoult, D. (2014). The origin and distribution of human lice in the world. Infection, Genetics And Evolution, 23, 209-217. https://doi.org/10.1016/j.meegid.2014.01.017

Boyd, B., & Reed, D. (2012). Taxonomy of lice and their endosymbiotic bacteria in the post-genomic era. Clinical Microbiology And Infection, 18(4), 324-331. https://doi.org/10.1111/j.1469-0691.2012.03782.x

Capinera, J. (2008). Encyclopedia of entomology. Springer.
Kittler, R., Kayser, M., & Stoneking, M. (2004). Molecular Evolution of Pediculus humanus and the Origin of Clothing. Current Biology, 14(24), 2309. https://doi.org/10.1016/j.cub.2004.12.024

Raoult, D., Reed, D., Dittmar, K., Kirchman, J., Rolain, J., Guillen, S., & Light, J. (2008). Molecular Identification of Lice from PreColumbian Mummies. The Journal Of Infectious Diseases, 197(4), 535-543. https://doi.org/10.1086/526520

Reed, D., Light, J., Allen, J., & Kirchman, J. (2007). Pair of lice lost or parasites regained: the evolutionary history of anthropoid primate lice. BMC Biology, 5(1). https://doi.org/10.1186/1741-7007-5-7

Toups, M., Kitchen, A., Light, J., & Reed, D. (2010). Origin of Clothing Lice Indicates Early Clothing Use by Anatomically Modern Humans in Africa. Molecular Biology And Evolution, 28(1), 29-32. https://doi.org/10.1093/molbev/msq234

Chapter 2

A. W. Bacot, "The Louse problem," Proceedings of the Royal Society of Medicine, vol. 10, pp. 61–94, 1917.

Casem, M. L. (2016). Cell Systems. Case Studies in Cell Biology, 345–371. https://doi.org/10.1016/b978-0-12-801394-6.00015-4

Centres for Disease Control and Prevention. (2019). Pubic "Crab" Lice. Retrieved May 16th, 2021, from, https://www.cdc.gov/parasites/lice/pubic/biology.html

Centres for Disease Control and Prevention (2019). Head Lice. Retrieved May 16th, 2021, from, https://www.cdc.gov/parasites/lice/head/disease.html#:~:text=It%20may%20take%204%E2%80%936,Irritability%20and%20sleeplessness

Weiss, R. A. (2009). Apes, lice and prehistory. Journal of Biology, 8(2), 20. https://doi.org/10.1186/jbiol114

Mumcuoglu K.Y. (2008) Human lice: Pediculus and Pthirus. In: Raoult D., Drancourt M. (eds) Paleomicrobiology. Springer, Berlin, Heidelberg. https://doi.org/10.1007/978-3-540-75855-6_13

Abdoul Karim Sangaré, Ogobara K. Doumbo, Didier Raoult. (2016). Management and Treatment of Human Lice. BioMed Research International. https://doi.org/10.1155/2016/8962685

The Editors of the Encyclopedia Britannica. (2009). Hans Zinsser. Retrieved May 16th, 2021, from, https://www.britannica.com/biography/Hans-Zinsser.

Weiss, R. A. (2009). Apes, lice and prehistory. Journal of Biology, 8(2), 20. https://doi.

org/10.1186/jbiol114

Lice Clinics of Texas. (2019). A Historical Look at Lice Over the Centuries. Retrieved May 16th, 2021, from, https://www.liceclinicsoftexas.com/a-historical-look-at-lice-over-the-centuries/#:~:text

Salgado, V. L. (1999). Resistant Target Sites and Insecticide Discovery. Pesticide Chemistry and Bioscience, 236–246. https://doi.org/10.1533/9781845698416.5.236

Sayyadi M, Sayyad S, & Vahabi A. (2014). Pediculosis Capitis: A Review Article. Life Science Journal, 11(3s), 26-30. Retrieved May 16th, 2021, from, http://www.lifesciencesite.com/lsj/life1103s/006_22645life1103s14_26_30.pdf

Stanford Education. (n.d.). Pediculosis. Retrieved May 16th, 2021, from, https://web.stanford.edu/group/parasites/ParaSites2002/pediculosis/pediculosis.html#:~:text=Pediculosis

Chapter 3

Centers for Disease Control and Prevention (2020). Typhus Fevers. Retrieved 12 May 2021, from https://www.cdc.gov/typhus/epidemic/index.html

Bechach, Y., Capo, C., Mege, J., & Raoult, D. (2008). Epidemic typhus. The Lancet, 8(7), 417-426. https://doi.org/10.1016/S1473-3099(08)70150-6

Kollipara, R., & Tyring, S.K. (2017). Tropical Dermatology (Second Edition).

Akram, S.M, Ladd, M., King, K.C. (2021). Rickettsia Prowazekii. StatPearls Publishing.

Raoult, D., Woodward, T., & Dumler J.S (2004). The history of epidemic typhus Infect Dis Clin North Am, 18, 127-140.

Zinsser, H. (1935). Rate, Lice, and history. Broadway house, London.

Gross, L. (1996). How Charles Nicolle of the Pasteur Institute discovered that epidemic typhus is transmitted by lice: Reminiscences from my years at the Pasteur Institute in Paris. Proc. Natl. Acad. Sci., 93, 10539-10540.

Weiss, E. (1988). Biology of Rickettsial Diseases.

Ming-yuan, F., Walker, D.H., Shu-rong, Y., & Qing-huai, L. (1987). Epidemiology and Ecology of Rickettsial Diseases in the People's Republic of China. Reviews of Infectious Diseases, 9(4), 823-840.

Conlon, J.M. The Historical Impact Of Epidemic Typhus.

Raoult D., Ndihokubwayo J.B., Tissot-Dupont H., Roux, V., Faugere, B., Abegbinni, R., & Birtles R.J. (1998). Outbreak of epidemic typhus associated with trench fever in Burundi. Lancet, 352(9125), 353-358. doi: 10.1016/s0140-6736(97)12433-3

Romero, A., Zeissig,O., España, D., Rizzo, L. (1977). Exanthematic typhus in Guatemala. Bol Oficina Sanit Panam, 83, 223-236.

Byam, W., Carroll, J.H, Churchill, J.H., et al. (1919). Trench fever, a louse-borne disease, Oxford University Press, London.

Bush, L.M, Vazquez-Pertejo, M.T. (2020). Trench Fever. Merck Sharp & Dohme Corp.

Boutellis, A., Abi-Rached, L., Raoult, D. (2014). The origin and distribution of human lice in the world. Infection, Genetics and Evolution, 23, 209-217. https://doi.org/10.1016/j.meegid.2014.01.017

Anstead, G.M. (2016). The centenary of the discovery of trench fever, an emerging infectious disease of World War 1. The Lancet, 16(8), 164-172. https://doi.org/10.1016/S1473-3099(16)30003-2

Badiaga, S., & Brouqui, P. (2012). Human louse-transmitted infectious diseases Clin. Microbiol. Infect., 18, 332-337.

Swift, H.F. (1920). Trench fever. Arch Intern Med (Chic), 26(1), 76-98. doi:10.1001/archinte.1920.00100010079006

Maurin, M., & Raoult D. (1996). Bartonella (Rochalimaea) quintana infections. Clin Microbiol Rev, 9, 273–92

Byam, W., & Lloyd, L. (1920). Trench Fever: Its Epidemiology and Endemiology. Proceedings of the Royal Society of Medicine, 13, 1-27. https://doi.org/10.1177/003591572001301501

Strong, R.P, Swift, H.F., Opie,E.L et al.(1918). Trench fever. Report of Commission. Medical Research Committee, Oxford University Press, American Red Cross, Oxford.

Badiaga, S., & Brouqui, P. (2012) Human louse-transmitted infectious diseases. Clin. Microbiol. Infect., 18, 332-337.

Brouqui, P. & Raoult, D. (2006). Arthropod-borne diseases in homeless Ann. NY Acad. Sci., 1078, 223-235.

Ohl, M.E., & Spach, D.H. (2000). Bartonella quintana and Urban Trench Fever. Clinical Infectious Diseases, 31(1), 131–135. https://doi.org/10.1086/313890

Larsson, C., Andersson, M., & Bergström, S. (2009). Current issues in relapsing fever. Current Opinion in Infectious Diseases, 22(5), 443-449. doi: 10.1097/QCO.0b013e32832fb22b

U.S. National Library of Medicine (2021). Relapsing fever. MedlinePlus. Retrieved 10 May 2021, from https://medlineplus.gov/ency/article/001350.htm

Southern, P.M., & Sanford, J.P. (1969). Relapsing Fever, A clinical and microbiological review. Medicine, 48(2), 129-145.

Bryceson, A.D., Parry, E.H., Perine, P.H., Warrell, D.A., Vukotich, D., & Leithead C.S. (1970). Louse-borne relapsing fever. Q J Med, 39(153), 129-170.

Raoult, D. & Roux, V. (1999). The body louse as a vector of reemerging human diseases. Clin Infect Dis. 29(4), 888-11. doi: 10.1086/520454.

Nordstrand A., Bunikis I, Larsson C, et al. (2007). Tickborne relapsing fever diagnosis obscured by malaria, Togo. Emerg Infect Dis, 13, 117–123.

Dworkin ,M.S, Schwan, T.G., & Anderson, D.E. (2002). Tick-borne relapsing fever in North America.
Med Clin North Am., 86(2):417-33

Chapter 4

Amanzougaghene, N., Mediannikov, O., Ly, T. D. A., Gautret, P., Davoust, B., Fenollar, F., & Izri, A. (2020). Molecular investigation and genetic diversity of Pediculus and Pthirus lice in France. Parasites and Vectors, 13(1). https://doi.org/10.1186/s13071-020-04036-y

Pietri, J. E., Yax, J. A., Agany, D. D. M., Gnimpieba, E. Z., & Sheele, J. M. (2020). Body lice and bed bug co-infestation in an emergency department patient, Ohio, USA. IDCases, 19. https://doi.org/10.1016/j.idcr.2020.e00696

Meister, L., & Ochsendorf, F. (2016, November 11). Head Lice. Deutsches Arzteblatt International. https://doi.org/10.3238/arztebl.2016.0763

Castelletti, N., & Barbarossa, M. V. (2020). Deterministic approaches for head lice infestations and treatments. Infectious Disease Modelling, 5, 386–404. https://doi.org/10.1016/j.idm.2020.05.002

Boumbanda-Koyo, C. S., Mediannikov, O., Amanzougaghene, N., Oyegue-Liabagui, S. L., Imboumi-Limoukou, R. K., Raoult, D., … Fenollar, F. (2020). Molecular identification of head lice collected in Franceville (Gabon) and their associated bacteria. Parasites and Vectors, 13(1). https://doi.org/10.1186/s13071-020-04293-x

Toups, M. A., Kitchen, A., Light, J. E., & Reed, D. L. (2011). Origin of clothing lice indicates early clothing use by anatomically modern humans in Africa. Molecular Biology and Evolution, 28(1), 29–32. https://doi.org/10.1093/molbev/msq234

Amanzougaghene, N., Fenollar, F., Raoult, D., & Mediannikov, O. (2020, January 21). Where Are We With Human Lice? A Review of the Current State of Knowledge. Frontiers in Cellular and Infection Microbiology. Frontiers Media S.A. https://doi.org/10.3389/fcimb.2019.00474

Lambiase, S., & Perotti, M. A. (2019). Using human head lice to unravel neglect and cause of death. Parasitology, 146(5), 678–684. https://doi.org/10.1017/S0031182018002007

Chapter 5

Durden, L. A., & Musser, G. G. (1994a). The sucking lice (Insecta: Anoplura) of the world: a taxonomic checklist with records of mammalian hosts and geographical distributions. Bulletin of the American Museum of Natural History. 218, 90.

Galloway, T. D. (2019). Phthiraptera of Canada. ZooKeys, (819), 301–310. https://doi.org/10.3897/zookeys.819.26160

Mullen, G. R., & Durden, L. A. (2019). Medical and veterinary entomology (3rd ed.). Elsevier Inc. https://doi.org/10.1016/B978-0-12-814043-7.00007-8

Palma, R. L., & Barker, S. C. (1996). Zoological Catalogue of Australia: Psocoptera, Phthiraptera, Thysanoptera (Vol. 26). CSIRO Publishing.

Price, R. D., Hellenthal, R. A., & Palma, R. L. (2003a). World checklist of chewing lice with host associations and keys to families and genera. Illinois Natural History Survey Special Publication, 24, 1–448.

Royal Entomological Society. (n.d.). Sucking and biting lice (Phthiraptera). https://www.royensoc.co.uk/entomology/orders/sucking-and-biting-lice#:~:text=-Sucking%20and%20biting%20lice%20(Phthiraptera)&text=Traditionally%20the%20group%20was%20divided,on%20both%20birds%20and%20mammals.

Smith, V. (2011, December 14). Phthiraptera. Phthiraptera.info. http://phthiraptera.info/taxonomy/term/24526/descriptions

Chapter 6

Brazier, Y. (2020, April 22). Pubic lice (crabs): Symptoms, risk factors, and treatment (C. Kay, Ed.). Retrieved May 13, 2021, from https://www.medicalnewstoday.com/articles/173681

Britannica, T. (2009, February 10). Sucking louse. Retrieved May 15, 2021, from https://www.britannica.com/animal/sucking-louse

Britannica, T. (2019, August 8). Pubic louse. Retrieved May 14, 2021, from https://www.britannica.com/animal/pubic-louse

Britannica, T. (2011, November 16). Chewing louse. Retrieved May 15, 2021, from https://www.britannica.com/animal/chewing-louse

CDC. (2019, June 26). CDC - lice - Pubic "CRAB" lice - biology. Retrieved May 14, 2021, from https://www.cdc.gov/parasites/lice/pubic/biology.html

CDC. (2019, October 15). CDC - lice - head lice - epidemiology & risk factors. Retrieved May 13, 2021, from https://www.cdc.gov/parasites/lice/head/epi.html

CDC. (2019, October 15). CDC - lice - head lice - treatment. Retrieved May 13, 2021, from https://www.cdc.gov/parasites/lice/head/treatment.html

CDC. (2019, September 11). CDC - Lice - Head Lice - Biology. Retrieved May 13, 2021, from https://www.cdc.gov/parasites/lice/head/biology.html

CDC. (2019, September 11). CDC - lice - head lice - Diagnosis. Retrieved May 13, 2021, from https://www.cdc.gov/parasites/lice/head/diagnosis.html

CDC. (2019, September 11). CDC - Lice - Head Lice - Disease. Retrieved May 13, 2021, from https://www.cdc.gov/parasites/lice/head/disease.html

CDC. (2019, September 12). CDC - lice - body lice - biology. Retrieved May 15, 2021, from https://www.cdc.gov/parasites/lice/body/biology.html

CDC. (2019, September 12). CDC - lice - body lice - Diagnosis. Retrieved May 15, 2021, from https://www.cdc.gov/parasites/lice/body/diagnosis.html

CDC. (2019, September 12). CDC - lice - body lice - disease. Retrieved May 15, 2021, from https://www.cdc.gov/parasites/lice/body/disease.html

CDC. (2019, September 12). CDC - lice - body lice - epidemiology & risk factors. Retrieved May 15, 2021, from https://www.cdc.gov/parasites/lice/body/epi.html

CDC. (2019, September 12). CDC - lice - body lice - Prevention & control. Retrieved May 15, 2021, from https://www.cdc.gov/parasites/lice/body/prevent.html

CDC. (2019, September 12). CDC - lice - body lice - treatment. Retrieved May 15, 2021, from https://www.cdc.gov/parasites/lice/body/treatment.html

CDC. (2019, September 12). CDC - lice - head lice - Prevention & control. Retrieved May 13, 2021, from https://www.cdc.gov/parasites/lice/head/prevent.html

CDC. (2019, September 12). CDC - lice - Pubic "CRAB" lice - Diagnosis. Retrieved May 14, 2021, from https://www.cdc.gov/parasites/lice/pubic/diagnosis.html

CDC. (2019, September 12). CDC - lice - Pubic "CRAB" lice - disease. Retrieved May 14, 2021, from https://www.cdc.gov/parasites/lice/pubic/disease.html

CDC. (2019, September 12). CDC - lice - Pubic "CRAB" lice - epidemiology & risk factors. Retrieved May 14, 2021, from https://www.cdc.gov/parasites/lice/pubic/epi.html

CDC. (2019, September 12). CDC - lice - Pubic "CRAB" lice - Prevention & control. Retrieved May 14, 2021, from https://www.cdc.gov/parasites/lice/pubic/prevent.html

CDC. (2019, September 12). CDC - lice - Pubic "CRAB" lice - treatment. Retrieved May 14, 2021, from https://www.cdc.gov/parasites/lice/pubic/treatment.html

CDC. (2020, August 31). CDC - lice - body lice - frequently Asked QUESTIONS (FAQS). Retrieved May 15, 2021, from https://www.cdc.gov/parasites/lice/body/gen_info/faqs.html
CDC. (2020, September 17). CDC - Lice - Head Lice - General Information - Frequently Asked Questions (FAQs). Retrieved May 13, 2021, from https://www.cdc.gov/parasites/lice/head/gen_info/faqs.html

CDC. (2020, September 17). CDC - lice - Pubic "crab" - general information - frequently asked Questions (FAQs). Retrieved May 14, 2021, from https://www.cdc.gov/parasites/lice/pubic/gen_info/faqs.html

Clay, T. (2017, November 2). Louse. Retrieved May 17, 2021, from https://www.britannica.com/animal/louse

Crabs. (2020, July 28). Retrieved May 13, 2021, from https://www.ashasexualhealth.org/crabs/

Guenther, L. C. (2021, March 24). Pediculosis and Phthiriasis (Lice Infestation). Retrieved May 13, 2021, from https://emedicine.medscape.com/article/225013-overview#a5

Healthline Editorial Team. (2019, September 11). Head Lice Infestation (C. Cobb, Ed.). Retrieved May 13, 2021, from https://www.healthline.com/health/head-lice

Mayo Clinic Staff. (2020, December 09). Body lice. Retrieved May 15, 2021, from https://www.mayoclinic.org/diseases-conditions/body-lice/diagnosis-treatment/drc-20350316

Mayo Clinic Staff. (2020, December 09). Body lice. Retrieved May 15, 2021, from https://www.mayoclinic.org/diseases-conditions/body-lice/symptoms-causes/syc-20350310

Mayo Clinic Staff. (2020, December 09). Pubic lice (crabs). Retrieved May 13, 2021, from https://www.mayoclinic.org/diseases-conditions/pubic-lice-crabs/diagnosis-treatment/drc-20350306

Mayo Clinic Staff. (2020, December 09). Pubic lice (crabs). Retrieved May 13, 2021, from https://www.mayoclinic.org/diseases-conditions/pubic-lice-crabs/symptoms-causes/syc-20350300

Mayo Clinic Staff. (2020, July 10). Head lice. Retrieved May 13, 2021, from https://www.mayoclinic.org/diseases-conditions/head-lice/diagnosis-treatment/drc-20356186
Mayo Clinic Staff. (2020, July 10). Head lice. Retrieved May 13, 2021, from https://www.mayoclinic.org/diseases-conditions/head-lice/symptoms-causes/syc-20356180

MSU Pesticide Safety Education Program. (2006). Questions and Answers About Head Lice [Pamphlet]. East Lansing, MI: Michigan State University.

Vafasso, J. (2019, March 8). Body Lice Infestation (C. Cobb, Ed.). Retrieved May 14, 2021, from https://www.healthline.com/health/body-lice
Winnipeg Regional Health Authority. (2008, January). Head Lice Life Cycle and Characteristics.

Chapter 7

Amanzougaghene, N., Fenollar, F., Raoult, D., & Mediannikov, O. (2019). Where Are We With Human Lice? A Review of the Current State of Knowledge (1376786973 1006084175 R. O. Rego, 1376786974 1006084175 C. J. Oliveira, & 1376786975 1006084175 S. J. Cutler, Eds.). Frontiers in Cellular and Infection Microbiology, 9(474). doi:https://doi.org/10.3389/fcimb.2019.00474

Branches. (n.d.). Retrieved May 16, 2021, from https://www.ebi.ac.uk/training/online/courses/introduction-to-phylogenetics/what-is-a-phylogeny/aspects-of-phylogenies/branches/

Carr, S. M. (2012). Concepts of monopoly, polyphyly, & paraphyly. Retrieved May 16, 2021, from https://www.mun.ca/biology/scarr/Taxon_types.htm

DeGrandpre, Z., MS, ND. (2017). What Are Lice, and Where Do They Come From? (1376731492 1006048483 J. Marcin M.D., Ed.). Retrieved May 16, 2021, from https://www.healthline.com/health/lice-what-are-lice

DNA Sequencing Fact Sheet. (n.d.). Retrieved May 16, 2021, from https://www.genome.gov/about-genomics/fact-sheets/DNA-Sequencing-Fact-Sheet

Dreyfuss, G., Philipson, L., & Mattaj, I. W. (1988). Ribonucleoprotein Particles in Cellular Processes. Journal of Cell Biology, 106(5), 1419-1425. doi:10.1083/jcb.106.5.1419.

El-Showk, S. (2015, September 28). Learning from Lice. Retrieved May 16, 2021, from https://www.nature.com/scitable/blog/accumulating-glitches/learning_from_lice/

GENOME SEQUENCING. (2003, January 15). Retrieved May 16, 2021, from http://www.genomenewsnetwork.org/resources/whats_a_genome/Chp2_1.shtml

Gibbons, A. (2012, June 13). Bonobos Join Chimps as Closest Human Relatives. Retrieved May 16, 2021, from https://www.sciencemag.org/news/2012/06/bonobos-join-chimps-closest-human-relatives

Harris, K. (2019, June 19). MRNA Splicing. Retrieved May 16, 2021, from https://bio.libretexts.org/Learning_Objects/Worksheets/Biology_Tutorials/mRNA_Splicing

Johnston, C., Martin, B., Fichant, G., Polard, P., & Claverys, J. (2014). Bacterial transformation: Distribution, shared mechanisms and divergent control. Nature Reviews Microbiology, 12, 181-196. Retrieved February 10, 2014, from https://www.nature.com/articles/nrmicro3199

Kelemen, O., Convertini, P., Zhang, Z., Wen, Y., Shen, M., Falaleeva, M., & Stamm, S. (2012). Function of alternative splicing. Gene, 514(1), 1-30. doi:10.1016/j.gene.2012.07.083
Kittler, R., Kayser, M., & Stoneking, M. (2003). Molecular Evolution of Pediculus humanus and the Origin of Clothing. Current Biology, 13(16), 1414-1417. doi:https://doi.org/10.1016/S0960-9822(03)00507-4

Lice DNA study shows humans first wore clothes 170,000 years ago. (2011, January 7). Retrieved May 16, 2021, from https://www.sciencedaily.com/releases/2011/01/110106164616.htm

Most, D., Ferguson, L., & Harris, R. A. (2014). RNA Splicing. Retrieved May 16, 2021, from https://www.sciencedirect.com/topics/neuroscience/rna-splicing#:~:text=RNA%20splicing%20is%20a%20process,of%20mRNA%20into%20a%20protein
Nickle, T., & Barrette-Ng, I. (n.d.). 11.3: Whole Genome Sequencing. Retrieved May 16, 2021, from https://bio.libretexts.org/Bookshelves/Genetics/Book%3A_Online_Open_Genetics_(Nickle_and_Barrette-Ng)/11%3A_Genomics_and_Systems_Biology/11.03%3A_Whole_Genome_Sequencing

O. (n.d.). Electron Transport Chain. Retrieved May 16, 2021, from https://courses.lumenlearning.com/wm-biology1/chapter/reading-electron-transport-chain/
Parey, K., Wirth, C., Vonck, J., & Zickermann, V. (2020). Respiratory complex I — structure, mechanism and evolution. Current Opinion in Structural Biology, 63, 1-9. doi:https://doi.org/10.1016/j.sbi.2020.01.004

Reed, D. L., Light, J. E., Allen, J. M., & Kirchman, J. J. (2007). Pair of lice lost or parasites regained: The evolutionary history of anthropoid primate lice. BMC Biology, 5(7). doi:https://doi.org/10.1186/1741-7007-5-7

Shizuya, H., & Kouros-Mehr, H. (2001). The development and applications of the bacterial artificial chromosome cloning system. The Keio Journal of Medicine, 50(1), 26-30. doi:10.2302/kjm.50.26

Torrent, D. (2011, January 1). Lice study dates first clothing at 170,000 years. Retrieved May 16, 2021, from https://www.floridamuseum.ufl.edu/science/lice-study-dates-first-clothing-at-170000-years/

Toups, M. A., Kitchen, A., Light, J. E., & Reed, D. L. (2011). Origin of Clothing Lice Indicates Early Clothing Use by Anatomically Modern Humans in Africa. Molecular Biology and Evolution, 28(1), 29-32. Retrieved May 16, 2021, from https://academic.oup.com/mbe/article/28/1/29/984822

Waterson, R. H., Lander, E. S., & Wilson, R. K. (2005). Initial sequence of the chimpanzee genome and comparison with the human genome. Nature, 437, 69-87. doi:https://doi.org/10.1038/nature04072

Zhang, C., Zhang, D., Zhu, T., & Yang, Z. (2011). Evaluation of a Bayesian Coalescent Method of Species Delimitation. Systematic Biology, 60(6), 747-761. Retrieved May 16, 2021, from https://www.jstor.org/stable/41316576?seq=1

Chapter 8

Barker, S. C., & Altman, P. M. (2010). A randomised, assessor blind, parallel group comparative efficacy trial of three products for the treatment of head lice in children--melaleuca oil and lavender oil, pyrethrins and piperonyl butoxide, and a "suffocation" product. BMC dermatology, 10, 6. https://doi.org/10.1186/1471-5945-10-6

Benelli, G., Caselli, A., Di Giuseppe, G., & Canale, A. (2018). Control of biting lice, MALLOPHAGA – a review. Acta Tropica, 177, 211-219. doi:10.1016/j.actatropica.2017.05.031

Birkemoe, T., Lindstedt, H. H., Ottesen, P., Soleng, A., Næss, Ø., & Rukke, B. A. (2016). Head lice predictors and infestation dynamics among primary school children in Norway. Family practice, 33(1), 23–29. https://doi.org/10.1093/fampra/cmv081

Canyon, D. V., & Speare, R. (2007). A comparison of botanical and synthetic substances commonly used to prevent head lice (Pediculus humanus var. capitis) infestation. International journal of dermatology, 46(4), 422–426. https://doi.org/10.1111/j.1365-4632.2007.03132.x

CDC. (2019, October 15). CDC - lice - head lice - epidemiology & risk factors. Retrieved May 10, 2021, from https://www.cdc.gov/parasites/lice/head/epi.html

CDC. (2020, September 17). CDC - lice - head lice - general information - TREAT-MENT FAQS. Retrieved May 10, 2021, from https://www.cdc.gov/parasites/lice/head/gen_info/faqs_treat.html#:~:text=Washing%2C%20soaking%2C%20or%20drying%20items,should%20be%20considered%20for%20cleaning.

CDC. (2020, September 17). CDC - lice - head lice - general information - frequently asked Questions (FAQs). Retrieved May 11, 2021, from https://www.cdc.gov/parasites/lice/head/gen_info/faqs.html#:~:text=Data%20show%20that%20head%20lice,do%20not%20kill%20head%20lice.

DSHS Texas. (n.d.). Head Lice Fact Sheet [PDF]. Texas: Texas Department of State Health Services. Health Service Region 1. Epidemiology.
Grant, D. I. (1989). Parasitic skin diseases in cats. Journal of Small Animal Practice, 30(4), 250-254. doi:10.1111/j.1748-5827.1989.tb01553.x

DeGrandpre, Z., MS, ND, & Marcin, J., MD. (2019, September 18). What are lice, and where do they come from? Retrieved May 11, 2021, from https://www.healthline.com/health/lice-what-are-lice
Guenther, L. C., MD, FRCPC, FAAD. (2021, March 24). Pediculosis and Pthiriasis (Lice Infestation). Retrieved May 10, 2021, from https://emedicine.medscape.com/article/225013-overview#a5

Irwin, P. J., & Jefferies, R. (2004). Arthropod-transmitted diseases of companion animals in Southeast Asia. Trends in Parasitology, 20(1), 27-34. doi:10.1016/j.pt.2003.11.00
Loyola University Health System. (2012, January 24). A parent's survival guide to lice. Retrieved May 10, 2021, from https://www.sciencedaily.com/releases/2012/01/120124134422.htm

Rossini, C., Castillo, L., & González, A. (2007). Plant extracts and their components as potential control agents against human head lice. Phytochemistry Reviews, 7(1), 51-63. doi:10.1007/s11101-006-9026-0

Sanders, D. P., & Stanhope, J. (n.d.). Human lice. Retrieved May 10, 2021, from https://extension.missouri.edu/g7394#life

Thomas, J. E., DVM. (2018, June). Lice of Dogs. Retrieved May 10, 2021, from https://www.merckvetmanual.com/dog-owners/skin-disorders-of-dogs/lice-of-dogs

Thomas, J. E., DVM. (2018, August). Lice of Cats. Retrieved May 10, 2021, from https://www.merckvetmanual.com/cat-owners/skin-disorders-of-cats/lice-of-cats

Washington State University. (n.d.). Do you have LOUSY Animals?: Animal Agriculture: Washington State University. Retrieved May 10, 2021, from https://extension.wsu.edu/animalag/content/do-you-have-lousy-animals/

Wondra, S. (2021, January 03). How to Get Rid of Chewing Lice on Pets. Chewing Lice Treatments. Retrieved May 11, 2021, from https://www.petcarerx.com/article/how-to-get-rid-of-chewing-lice-on-pets/1563#:~:text=Unlike%20ticks%20and%20fleas%2C%20they,the%20base%20of%20the%20fur

Chapter 9

Anderson, A. L., & Chaney, E. (2009). Pubic lice (Pthirus pubis): history, biology and treatment vs. knowledge and beliefs of US college students. International journal of environmental research and public health, 6(2), 592-600.

Bechah, Y., Capo, C., Mege, J., & Raoult, D. (2008). Epidemic typhus. The Lancet Infectious Diseases, 8(7), 417–426. https://doi.org/10.1016/S1473-3099(08)70150-6

De Liberato, C., Magliano, A., Romiti, F., Menegon, M., Mancini, F., Ciervo, A., Di Luca, M., & Toma, L. (2019). Report of the human body louse (Pediculus humanus) from clothes sold in a market in central Italy. Parasites & Vectors, 12(1), 201–201. https://doi.org/10.1186/s13071-019-3458-z

Fournier, P., Ndihokubwayo, J., Guidran, J., Kelly, P., & Raoult, D. (2002). Human pathogens in body and head lice. Emerging Infectious Diseases, 8(12), 1515–1518. https://doi.org/10.3201/eid0812.020111

Hemond, C. (2012). The nitty-gritty on nits. Canadian Medical Association Journal (CMAJ), 184(9), E463–E464. https://doi.org/10.1503/cmaj.109-4173

Hurst, S., Dotson, J., Butterfield, P., Corbett, C., & Oneal, G. (2020). Stigma resulting from head lice infestation: A concept analysis and implications for public health. Nursing Forum (Hillsdale), 55(2), 252–258. https://doi.org/10.1111/nuf.12423

Nash, B. (2003). Treating head lice. Bmj, 326(7401), 1256-1257.

Portillo, A., Santibáñez, S., García-Álvarez, L., Palomar, A., & Oteo, J. (2015). Rickettsioses in Europe. Microbes and Infection, 17(11-12), 834–838. https://doi. org/10.1016/j.micinf.2015.09.009

Silva, L., Alencar, R. D. A., & Madeira, N. G. (2008). Survey assessment of parental perceptions regarding head lice. International journal of dermatology, 47(3), 249-255.

Weissmann, G. (2005). Rats, lice, and zinsser. Emerging Infectious Diseases, 11(3), 492–496. https://doi.org/10.3201/eid1103.AD1103

Chapter 10

Amanzougaghene, N., Fenollar, F., Raoult, D., & Mediannikov, O. (2020). Where Are We With
Human Lice? A Review of the Current State of Knowledge. Frontiers in Cellular and Infection Microbiology, 9,474. doi: 10.3389/fcimb.2019.00474

Candy, K., Nicolas, P., Andriantsoanirina, V., Izri, A., & Durand, R. (2017). In vitro efficacy of
five essential oils against Pediculus humanus capitis. Parasitol Res, 117(2),603-609. Retrieved from https://pubmed.ncbi.nlm.nih.gov/29264717/

Carroll, A. What are sea lice? Alaska Department of Fish and Game. Retrieved from https://www.adfg.alaska.gov/index.cfm?adfg=wildlifenews.view_article&articles_id=388

Galassi, F., Ortega-Insaurralde, I., Adjeman, V., Gonzales-Audino, P., Picollo, M., & Toloza, A.
(2021). Head lice were also affected by COVID-19: a decrease on Pediculosis infestation during lockdown in Buenos Aires. Nature Public Health Emergency Collection, 1-8. Retrieved from https://www.ncbi.nlm.nih.gov/pmc/articles/PMC7787699/

Government of Canada. (2019). Scientific research on sea lice. Retrieved from https://www.dfo-mpo.gc.ca/aquaculture/sci-res/species-especes/sea-lice-poux-eng.htm

Greive, K., & Barnes, T. (2017). The efficacy of Australian essential oils for the treatment of head lice infestation in children: A randomised controlled trial. The Australian Journal of Dermatology, 59(2),99-105. Retrieved from https://www.ncbi.nlm.nih.gov/pmc/articles/PMC6001441/

Weintraub, K. (2017). Revenge of the Super Lice. Scientific American. Retrieved from https://www.scientificamerican.com/article/revenge-of-the-super-lice/

Willingham, E. (2011). Of lice and men: An itchy history. Scientific American. Retrieved from https://blogs.scientificamerican.com/guest-blog/of-lice-and-men-an-itchy-history/